NANOTECHNOLOGY
A LEARNING GUIDE

ABOUT THE AUTHOR

Dr. M. Prasanthrajan, is presently serving as Head, Centre for Agricultural Nanotechnology, Tamil Nadu Agricultural University, Coimbatore, Tamil Nadu, India. His teaching experience with the Under Graduate, Post Graduate and Doctoral courses in the field of Nanotechnology helped him to write this book for easy understanding of the subject. His post doctoral training in the field of nanoremediation at the University of California, Riverside, USA, during his early career set his goal in publishing a book on nanotechnology. He has published 61 research papers in peer reviewed National and International journals. His books "Objective Environmental Studies and Disaster Management" and "Objective Environmental Pollution and Climate Change" were well received by the students of India. He has obtained fourteen prestigious awards including two international travel grants. He has visited USA, Singapore, Thailand, Indonesia and UAE to share his expertise in the field of Environmental Science and Nanotechnology to the international arena.

Dr. A.S. Goveanthan worked at TNAU as Senior Research fellow at Centre for Agricultural Nanotechnology and worked as Research Associate at Water Technology Centre. He obtained his M.Sc. degree in Environmental Science in the year 2016 at TNAU. He got his Ph.D. degree in Environmental Science in the year 2020 at TNAU. He is a gold medalist and got best thesis award for his Doctoral research work. He had cleared ASRB NET examination. He handled classes for M.Sc. and Ph.D. students of Nano Technology. He had published various National, International papers, book and book chapters.

Dr. A. Lakshmanan is one of the leading Nanotechnologists in the country. Professor Lakshmanan served as Head, Centre for Agricultural Nanotechnology during 2019 to 2022, later he served as the Dean, School of Post Graduate Studies, Tamil Nadu Agricultural University. He has published more than 80 research papers in the peer reviewed national and international journals. He has authored five books. His book "Insights into Fundamentals & Applications of nanotechnology" has well reached with the students. He has three patents and visited many countries like USA, Canada, Mexico, Singapore, Norway, UK, Sweden, Netherlands, Srilanka, Kenya, Switzerland and Philippines to share his research experience.

Ms. C. Indhu Parameshwari, Scholar of the Department of Environmental Science, TNAU, Coimbatore. She completed her UG degree in Horticulture during the year 2016 at Horticultural College and Research Institute, Periyakulam. She completed her M.Sc. degree in Environmental Science in the year 2018 at TNAU. She had cleared UGC NET, ASRB NET and appeared for ARS interview. She had published various National, International papers and book chapters connected with nanotechnology.

Dr. S. Shalini, has a doctorate in Chemistry. She has extensively worked on the Arsenic removal from ground water using Pectin stabilized nano zero valent iron. Presently she is working as visiting faculty and taking classes for the Under Graduate and Post Graduate students in the field of Chemistry and Nanochemistry. She has published more than 10 research papers in high impact factor journals besides attending many conferences.

Dr R. Abhinayaa is working as a Teaching assistant; she did her DST-SERB National postdoctoral fellowship in the Centre for Agricultural Nanotechnology, Tamil Nadu Agricultural University. She has an expertise in the field of environmental nanotechnology and nanotoxicology and have handled classes for M.Sc. and Ph.D. students and her SERB project also relies on developing nanotechnology based eco-friendly and economically viable tools to capture Greenhouse gasses.

NANOTECHNOLOGY
A LEARNING GUIDE

Dr. M. Prasanthrajan, Ph.D.
Professor and Head
Centre for Agricultural Nanotechnology
Tamil Nadu Agricultural University
Coimbatore – 641 003
Tamil Nadu

Dr. A.S. Goveanthan, Ph.D.
Senior Research Fellow
Centre for Agricultural Nanotechnology
Tamil Nadu Agricultural University
Coimbatore – 641 003
Tamil Nadu

Dr. A. Lakshmanan, Ph.D.
Former Dean
School of Post Graduate Studies
Tamil Nadu Agricultural University
Coimbatore – 641 003
Tamil Nadu

Ms. C. Indhu Parameshwari, M.Sc.
Assistant Professor
Department of Agricultural Engineering
Easa College of Engineering and Technology
Palakkad Main Road
Navakkarai Post
Coimbatore – 641 105
Tamil Nadu

Dr. S. Shalini, Ph.D.
PG Teacher
SBKV Higher Secondary School
RS Puram, Coimbatore – 641 002
Tamil Nadu

Dr. R. Abhinayaa, Ph.D.
Teaching Assistant
School of Post Graduate Studies
Tamil Nadu Agricultural University
Coimbatore – 641 003
Tamil Nadu

SCIENTIFIC® PUBLISHERS
New Delhi • Jodhpur

SCIENTIFIC® PUBLISHERS

Published by:

Jodhpur
5 A, New Pali Road, P.O. Box 91
Jodhpur - 342 001 Rajasthan (India)
E-mail: info@scientificpub.com
Website: www.scientificpubonline.com

New Delhi
4806/24, Ansari Road, Daryaganj,
New Delhi - 110 002 (India)

ISBN: 978-81-96246-73-0
eISBN: 978-81-96246-71-6

Printed in India

Content

1

MCQs Nanotechnology

1. One nm is equal to ___________
   (a) 10^{-6} cm (b) 10^{9} cm
   (c) 10^{9} µm (d) 10^{-9} m
2. The prefix "nano" comes from a
   (a) French word meaning billion (b) Spanish word meaning particle
   (c) Greek word meaning dwarf (d) Latin word meaning invisible
3. Who first introduced the concept of nanotechnology?
   (a) Richard Smalley (b) Eric Drexier
   (c) Sumio lijima (d) Richard Feynmann
4. Buckyball is a
   (a) A carbon molecule
   (b) Plastic explosive nanoparticle
   (c) A futuristic concept car (C111)
   (d) Concrete nanoparticle with strength of 20 nanonewtons
5. Buckyball was discovered by
   (a) Alexei Ekimove (b) Louis Brus
   (c) Eric Drexier (d) RobURT curl
6. Which is an allotrope of carbon?
   (a) Graphite (b) Diamond
   (c) Fullerenes (d) All the above
7. Father of modern nanotechnology
   (a) Norio Taniguchi (b) Eric Drexier
   (c) Richard Feynmann (d) Richard Smalley
8. Carbon nanotube was discovered by
   (a) Norio Taniguchi (b) Sumio lijima
   (c) Richard Feynmann (d) Richard Smalley
9. According to Moore's law, the amount of space required to install a transistor on a chip shrink roughly by half for
   (a) Every 26 months (b) Every 18 months
   (c) Every 28 months (d) Every 19 months
10. The lecture titled "There is plenty of rooms at bottom" was delivered by
    (a) Norio Taniguchi (b) Eric Drexier
    (c) Richard Feynmann (d) Richard Smalley

11. Electron microscope was invented by
    (a) Ernst Ruska and Max Knoll (b) Binning and Roher
    (c) Alloissenfelder (d) Kroto
12. Atomic force microscope was invented by
    (a) Gerd Binnig, Calvin quate and Christoph Gerber
    (b) Binning and Roher
    (c) Alloissenfelder (d) Ernst Ruska and Max Knoll
13. Atomic force microscope was invented Scanning tunneling microscope was invented in the year
    (a) 1987 (b) 1985
    (c) 1986 (d) 1988
14. Who discovered colloidal "ruby" gold
    (a) Michael Faradey (b) Binning and Roher
    (c) Alexel Ekimov (d) Ernst Ruska and Max Knoll
15. Zero dimensional nanomaterials
    (a) Nanotubes (b) Nanowires
    (c) Nano Particles (d) Nano films
16. The term quantum dot was coined by
    (a) Alexei Ekimove (b) Louis Brus
    (c) Mark Reed (d) Roburt curl
17. Quantum dot was discovered in
    (a) 1979 (b) 1980
    (c) 1981 (d) 1982
18. Quantum dot was discovered by
    (a) Alexei Ekimove (b) Binning and Roher
    (c) Mark Reed (d) Roburt curl
19. N-type semiconductors are doped with
    (a) Boron (b) Germanium
    (c) Antimony (d) Bismuth
20. P- type semiconductors are doped with
    (a) Boron (b) Arsenic
    (c) Antimony (d) phosphorus
21. Physical treatment of wastewater involves the processes like
    (a) Coagulation (b) sedimentation
    (c) Flocculation (d) All of the above
22. Application of ____________ to remove the pathogenic microorganisms is a chemical treatment
    (a) Chloride powder (b) Chlorine powder
    (c) Acidic Acid (d) Elite Microbes
23. Nano filtration is a promising approach in water purification that immobilizes nano-materials on or within a solid matrix called

(a) Nano membrane (b) Nano filtration
(c) Mixed matrix membranes (d) Nano composites based membranes

24. Mixed matrix membranes consist of continuous polymer phase in which the nano-material like ____________ are incorporated
(a) Zeolites (b) carbon nanotubes
(c) Carbon molecular sieve (d) All the above

25. Mixed matrix membranes are divided into ________ broad categories based on the filler materials used in the dispersed phase.
(a) Four (b) Six
(c) Three (d) Seven

26. Zeolites are naturally occurring ___________ minerals with highly uniform sub-nanometer and nanometer scale crystalline structures.
(a) Sub nanometer (b) Nanometer scale
(c) Aluminosilicate (d) cellulose acetate

27. ________ based nanomembranes are currently developing in the water purification due to the water permeability on the range of ultrafiltration with the solute selectivity
(a) Kaolinite (b) Montmorillonite
(c) Lllite (d) Zeolite

28. Zeolite nano-particles impregnated polysulfone membranes enhanced water hydraulic permeability and metal exchange capacity of the membrane and the maximum sorption capacity of the hybrid membrane for the ____________
(a) Lead (b) Nickel
(c) Lithium (d) Both A and B

29. __________ is a reservoir of abundant functional groups in the surface, which serves as active sites of binding the pollutants and also that can be tailored for excellent selectivity towards specific pollutant
(a) Silica (b) Zinc
(c) Iron (d) copper

30. _________ hold promise for water filtration
(a) Crystalline silica (b) Silica sand
(c) Green silica (d) Quartz

31. Inorganic nanofiltration consists of thin inorganic nanomembrane which are mixed oxides from ___________ over a porous ceramic support layer which provides good mechanical stability
(a) Aluminium (Al) (b) Zirconium (Zr)
(c) Titanium (Ti) (d) All the Above

32. Cellulose coated magnetite-silver nanoparticles antimicrobial properties that can be used against ______ and it could be used effectively for water purification
(a) Clostridium (b) Shigella
(c) Salmonella (d) E. coli

33. Nano-hybrid membrane contains ____________ has excellent antibacterial property
    (a) Polymer (b) chitosan
    (c) Carbon nanotube (d) Rhodamine
34. ____________ ultrafiltration membranes in combination with graphene oxide and HKUST-1 (a metal–organic framework) exhibits higher hydrophilicity, antifouling property and highest water permeate flux
    (a) Biodegradable cellulose acetate
    (b) Phosphonate cellulose acetate
    (c) Bio-inspired membranes
    (d) Biomimetic membranes
35. ____________ are also called bio-inspired membranes are the synthetic membranes
    (a) Nanofibrous membranes (b) Biomimetic membranes
    (c) Cellulose membranes (d) Nanocomposite
36. Reduced graphene oxide has tremendous application in __________ through the interlayer spacing.
    (a) Drying (b) Disinfection
    (c) Desalination (d) Chlorination
37. ____________ membranes in the presence of candle soot carbon nanoparticles contributed good percentage of salt rejection with a slight reduction in the water flux behavior could be used for water desalination
    (a) Cellulose acetate (b) Polyethylene glycol
    (c) Chitosan (d) Both A and B
38. ______________ based electrosorption is an environmental friendly desalination technique compared to other conventional technologies of membrane separation and thermal distillation
    (a) Capacitive deionization (b) Porous Electron
    (c) Graphene (d) Nanoparticle hybrid
39. ______________ of saline water using Nano composites is an emerging trend in the desalination process
    (a) Reverse osmosis (b) Semipermeable
    (c) Capacitive deionization (d) solar stills
40. Zeolites has strong affinity for ______________
    (a) Carbon dioxide (b) Nitrous oxide
    (c) Carbon monoxide (d) Sulfur dioxide
41. When the world renowned Theoretical Physicist Dr. Richard Feynman delivered a classic lecture on
    (a) December 29th ,1959 (b) January 13th , 1958
    (c) March 22nd, 1959 (d) December 20th, 1959
42. The term "nanotechnology" was in use as early as
    (a) 1977 (b) 1972
    (c) 1980 (d) 1974

43. who placed his nanotechnology focus on molecular machining saying that the "new technology will handle individual atoms and molecules with control and precision"__________
    (a) Drexler (b) Taniguchi
    (c) Ashrafi (d) Dr. Richard Feynman
44. Classification of solid based on the atomic arrangements are
    (a) Crystalline, Amorphous (b) Polycrystalline, Crystalline
    (c) Amorphous (d) All of the above
45. Particle theory is the basic model to explain the properties and behaviors in three states namely
    (a) Solids (b) liquids
    (c) gases (d) All the above
46. ______ are rigid structures but they lack a well-defined shape
    (a) Crystalline solids (b) Polycrystalline solids
    (c) Amorphous solids (d) All of the above
47. The most common example of an amorphous solid is ____________
    (a) Glass (b) Sterile soil
    (c) Sol gel (d) All of the above
48. Points in space about which these atoms are located such points in space are called as ____
    (a) lattice points (b) Crystal
    (c) Unit Cell (d) Primitive Cell
49. ____________ refers to the geometry of a set of points in space.
    (a) Crystal lattice (b) Rotation
    (c) Reflection (d) Inversion
50. __________ defined as a regular, indefinitely repeated array of points in three dimensions in which the points lie at the intersections of three sets of parallel equidistant planes
    (a) Space lattice (b) Crystal lattice
    (c) Lattice points (d) Both A and B
51. The unit assembly of atoms or molecules identical in composition, arrangement and orientation is called _______
    (a) Pattern (b) Basis
    (c) Crystal (d) Both A and B
52. ________ is the structural and functional unit of a crystal
    (a) Unit cell (b) Primitive cell
    (c) Multi cell (d) Both A and B
53. _______ can be a unit cell
    (a) Unit cell (b) Primitive cell
    (c) Mono cell (d) Both A and B
54. Unit cell has a geometry known as ________
    (a) Parallele piped (b) Parallelograms
    (c) Primitive centering (d) Single phase

55. Primitive unit cells contain only _________ which is made up from the lattice points at each of the corners
    (a) One lattice point (b) Two lattice point
    (c) Six lattice point (d) Nine lattice point
56. The two-dimensional primitive cells are
    (a) Parallelograms (b) Rhombus
    (c) Orthorhombic (d) Parallelepiped
57. A three-dimensional primitive cell is known to be
    (a) Parallelepiped (b) Parallelograms
    (c) Orthorhombic (d) Both A and B
58. Types of three-dimensional primitive cells are___________
    (a) Parallelepiped (b) Oblique rhombic prism
    (c) Oblique rectangular prism (d) All of the above
59. Crystal systems classified into ___________
    (a) Seven groups (b) Six groups
    (c) Five groups (d) Two groups
60. Basic crystal systems are ___________
    (a) Cubic (b) Tetragonal
    (c) Orthorhombic (d) All of the above
61. _________ are defined as two of the crystal axes are perpendicular to each other, but the third is obliquely incline
    (a) Monoclinic Crystal Systems (b) Triclinic Crystal Systems
    (c) Orthorhombic Crystal Systems(d) Hexagonal Crystal Systems
62. _________defined as none of the crystal axes is perpendicular to any of the others and the repetitive intervals are different along all the three axes.
    (a) Cube Crystal Systems (b) Triclinic Crystal Systems
    (c) Tetragonal Crystal Systems (d) Trigonal Crystal Systems
63. Trigonal sometimes called ____ crystal systems are defined as the three axes are equal in length and are equally inclined to each other at an angle other than.
    (a) Rhombohedral (b) Orthorhombic
    (c) Rectangular cuboid (d) Tetragonal
64. Centre of symmetry is also known as ___________
    (a) Inversion centre (b) Pairs of parallel
    (c) Opposite faces (d) Symmetrical positions
65. Any deviation from the perfect atomic arrangement in a crystal is said to contain ___________
    (a) Imperfections (b) Defects
    (c) Fine (d) Both A and B
66. Conductivity of some semiconductors is due entirely to trace amount of __________
    (a) Chemical impurities (b) Physical impurities
    (c) Biological impurities (d) Both A and B

67. Imperfections in crystalline solids are normally classified according to their dimension as
    (a) Point imperfections (b) Line imperfections
    (c) Surface imperfections (d) All of the above
68. Defects at some points in the crystal are also called____________
    (a) Zero dimensional defects (b) Two dimensional defects
    (c) Four dimensional defects (d) Five dimensional defects
69. Some atoms may not be present in their regular atomic sites ____________
    (a) Lattice site defects (b) Schottky defect
    (c) Frenkel defect (d) All of the above
70. At a lattice point, one or two or three atoms are missed, and this is referred to as ____________
    (a) Single vacancies (b) Triple vacancies
    (c) Two vacancies (d) All the above
71. In ionic crystals, if a cation vacancy exists, then in the very nearby place an anion vacancy also exists is called ____________
    (a) Scottky defect (b) Electronic defects
    (c) Compositional defect (d) Interstitial defect
72. If an atom is moved to an interstitial space in the crystal, then the defect is known as ____________
    (a) Scottky defect (b) Electronic defects
    (c) Interstitial defect (d) Line defects
73. In ionic crystals, if a cation (positive ion) moves to an interstitial space, then a vacancy is formed in its atomic position, then the defect is known as ____________
    (a) Scottky defect (b) Electronic defect
    (c) Compositional defect (d) Frenkel defect
74. The presence of impurity atoms in the crystal leads to ____________
    (a) Interstitial defect (b) Compositional defects
    (c) Electronic defects (d) Lattice site defects
75. During the crystallization few foreign atoms occupy the regular parental atom sites is known as ____________
    (a) Interstitial impurity defect (b) Substitutional defect
    (c) Compositional defect (d) Interstitial defect
76. The spaces between the parental atoms in a crystal are known as ____________
    (a) Germanium atoms (b) Interstitial atoms
    (c) Interstitial spaces (d) Interstitial defect
77. Defects formed due to the presence of the interstitial atoms are known as ____________
    (a) Interstitial spaces (b) Interstitial defect
    (c) Interstitial impurity defect (d) Substitutional defect
78. Non uniformity of charge or energy distribution in the crystal is referred to as ____________

(a) Electronic defect (b) Line Defects
(c) Edge Dislocation (d) Crystal surface

79. A crystal plane ends somewhere in the crystal, then along the edges of that incomplete plane produces defect in a crystal called ____________
(a) Line defect (b) Screw Dislocation
(c) Grain Boundaries (d) Stacking Fault

80. ___________ arise from a change in the stacking of atomic planes across a boundary.
(a) Screw Dislocation (b) Surface imperfections
(c) Edge Dislocation (d) Interstitial impurity

81. ___________ are those surface imperfections which separate crystals of different orientation in a poly-crystalline aggregation during nucleation
(a) Tilt boundaries (b) Grain boundaries
(c) Twin boundaries (d) Self-Interstitial

82. _________ is a surface imperfection that results from the stacking of one atomic plane out of sequence on another while the lattice on either side of the fault is perfect
(a) Schottky (b) Voids
(c) Stacking fault (d) Inclusions

83. A small angle grain boundary is formed when the edge dislocations are aligned in the manner is called as _____________
(a) Tilt boundary (b) Grain boundaries
(c) Twin boundaries (d) Three-dimensional

84. __________ are the boundaries in the grains at which the atomic arrangements on one side of the boundary is the mirror image of the atoms on the other side
(a) Tilt boundary (b) Grain boundaries
(c) Twin boundaries (d) Three-dimensional

85. Three-dimensional macroscopic defects are called ____________
(a) Line Defects (b) Compositional defect
(c) Interstitial defect (d) Bulk defects

86. Bragg's Law derived by the English physicists ____________in 1913 to explain why the cleavage faces of crystals appear to reflect X-ray beams at certain angles of incidence (theta, θ).
(a) Sir W.H. Bragg (b) W.L.Bragg
(c) Paul Bragg (d) Lawrence Bragg

87. Bragg's Law refers to _____________
(a) ∑k=1nVk=0 (b) λ = AB +BC
(c) nλ = 2d sin θ (d) c =1/(√ μ0 ε0)

88. _____________ is a formula that relates the size of sub micrometer crystallites, in a solid to the broadening of a peak in a diffraction pattern.
(a) Differential equation (b) Homogeneous differential equation
(c) Scherrer equation (d) Linear differential equation

89. The Scherrer equation can be written as ____________
(a) $\lambda = h/momentum = h/mv = h/p$ (b) $\sigma = nZe\mu$
(c) $\tau = K\lambda/\beta Cos\theta$ (d) $n\lambda = 2d \sin \theta$

90. ____________ are formed by the self-organization or chemical binding of the several organic molecules.
(a) Inorganic Nano materials (b) Organic Nano materials
(c) Composite Nano materials (d) Carbon Nano materials

91. ____________ are key components in the fabrication of organic nanoparticles with polar and non-polar regions aids in the encapsulation process.
(a) Amino acids (b) Amine group
(c) Basic (d) Zwitter ions

92. Organic nanomaterials are ____________
(a) Carbon nanotubes (b) Fullerenes
(c) Graphene nanomaterials (d) All of the above

93. ____________ are efficient in the remediation of persistent pollutants in the environment.
(a) Inorganic Nano materials (b) Organic Nano materials
(c) Composite Nano materials (d) Carbon Nano materials

94. Carbon nanotubes are allotropes of carbon with a cylindrical nanostructure which has length-to-diameter ratio of up to ____________
(a) 132,000,000:1 (b) 132,000:1
(c) 132,000,000,0000:1 (d) 132 :1

95. CNTs has ____________
(a) High mechanical strength (b) High electrical conductivity
(c) High ductility (d) All of the above

96. Carbon nanotube classified as ____________
(a) Multi-walled carbon nanotubes (b) Single-walled carbon nanotube
(c) Both A and B (d) None of the above

97. Carbon nanotube classified based on the layer of ____________ present in the nanotubes
(a) Graphite sheet (b) Graphene sheets
(c) PVC sheet (d) Flannel

98. Multi-walled carbon nanotubes (MWCNTs) promote____________
(a) Heavy metal sedimentation (b) Microbial activity
(c) Enzymes Activity (d) Aqueous solutions

99. Multi-walled carbon nanotubes (MWCNTs) promote ____________ in water–sediment interface
(a) Hardness (b) Heavy metal immobilization
(c) Aromaticity (d) Hydrophobicity

100. ____________ alleviate the bioavailability and transformation ability of the metals.

(a) Multi-walled carbon nanotubes (b) Single-walled carbon nanotube
(c) Carbon nanotube (d) Both A and B

101. ____________ with its high surface area, geometry and chemical properties used as promising sorbent for organic and inorganic pollutants in the wastewater treatment
(a) Graphene (b) Chitosan
(c) Graphene oxide (d) Both A and B

102. Toxic organic pollutants like dyes, brominated flame retardants, pharmaceuticals, endocrine disrupting chemicals and polycyclic aromatic hydrocarbons in the wastewater are removed efficiently with the ____________
(a) Graphene-based adsorbents (b) Graphene-based absorbents
(c) Graphene-based desorption (d) Both A and B

103. ________ has excellent adsorption-desorption rate, in wastewater treatment.
(a) Graphene (b) Fullerenes
(c) Graphite (d) Carbon

104. Fullerene and its composites have been studied widely for sorption of ____________
(a) Microbes (b) Salts
(c) Heavy metals (d) Gases

105. _________carbon composite harvested from the pyrolysis of waste biomass could be effectively used to store electrochemical energy.
(a) Nano-fibers (b) Mesoporous
(c) Nano tubes (d) Both A and B

106. Carbon nanofibers film prepared from the coal liquefaction residue by the electrospinning process found to be effective in ____________
(a) Coal waste management (b) Oil waste management
(c) Solid waste management (d) Nuclear waste management

107. Chitosan based nanomaterials are ____________
(a) Nano gels (b) Nano composites
(c) Nano formulations (d) Both A and B

108. Chitosan based nanomaterials serves as ____________
(a) Moisture absorbing agent (b) Moisture retention agent
(c) Moisture releasing agent (d) Moisture adsorption agent

109. __________is an aluminosilicate clay mineral made up of aluminium, silicon and hydrogen.
(a) Chitosan (b) Graphene
(c) Fullerenes (d) Halloysite

110. Halloysites are nanotubes of__________ in length
(a) 500–1000 nm (b) 50 - 1000 nm
(c) 500 nm – 1000 nm (d) 5 nm – 1000 nm

111. Halloysites are nanotubes of__________ in inner-diameter
(a) 1–100 nm (b) 10–100 nm
(c) 10–1000 nm (d) 1–1000 nm

112. _____ helps in degradation of air pollutants such as Nitric Oxide
(a) MoS_2 (b) CdS
(c) TiO_2 (d) Fe

113. Ag nanoparticle acts as _______________
(a) Water disinfectant (b) Antimicrobial agents
(c) Dye degradation (d) All of the above

114. Zn nanoparticles induces ____________
(a) Cyanide degradation (b) Photo catalysts
(c) Water treatment (d) All of the above

115. Cu nanoparticles involves in Degradation of aromatic compounds _______________
(a) 2-nitrophenol (b) 4-nitrophenol
(c) Organic dye (d) Both A and B

116. Pd nanoparticles helps in__________ removal
(a) Nitro aromatics (b) Diatrizoate
(c) Biosensors (d) Nitrate

117. Au nanoparticles helps in _____________ reduction in water
(a) Diatrizoate (b) Chromate
(c) Nitro aromatics (d) Trichloroethylene

118. The excellent removal mechanism of Cu2+ and Co2+ ions from wastewater was achieved using _________ by adsorption and redox reactions respectively
(a) nzvi (b) Zn
(c) Ag (d) Cu

119. nZVI helps in removal of_______ from ground water
(a) Cd (II) (b) Pb(II)
(c) As (III) (d) Sb (III)

120. __________ are the crystalline polymers engrossing structural design which is capable of assembling organic bridging linkers or organic ligands and metal ions or metal clusters such as lanthanide metals or transition metals
(a) CNT's (b) MOFs
(c) MIOFs (d) CBNP

121. MOFs in removal of _________ from wastewater
(a) Organic Matter (b) Microbes
(c) Heavy metals (d) Impurities

122. Ceramic Nano crystalline materials are solids with grain size of lower than ____________
(a) 1 nanometers (b) 10 nanometers
(c) 100 nanometers (d) 1000 nanometers

123. _______ are catalysts used in the petroleum refining industry and chemical production industries.
(a) Vermiculite (b) Kaolinite
(c) Zeolites (d) Montmorillonite

124. _________ has selective catalytic reduction of NOx
(a) Nano porous candidates (b) Faujasite zeolites
(c) Nano clays (d) Nano zeolite

125. Zeolites with crystal sizes less than ________ have a higher adsorption capacity
(a) 100 nm (b) 1 nm
(c) 30 nm (d) 180 nm

126. ___________ increases the water use efficiency as well as alter CH_4 producing (Methanogens).
(a) Nano clays (b) Nano zeolite
(c) Vermiculite (d) Both A and B

127. Nano clay and Nano eolite usage may be used as a strategy to achieve the water use efficiency and _____ reduction in the paddy field
(a) Nitrous emission (b) Methane emission
(c) Sulfur emission (d) CO_2 emission

128. Nano clay and Nano zeolite applied field generally requires _________ of water than conventional irrigation system.
(a) More amount (b) Mega amount
(c) Less amount (d) Low amount

129. Three nanoparticles studied were zinc silica nanogel liquid, copper silica nanogel liquid, and n-acetylcystein liquid coated zinc oxide quantum dots
(a) zinc silica (b) copper silica
(c) n-acetylcystein (d) All the above

130. Three types of nanoparticles such as zinc silica nanogel liquid, copper silica nanogel liquid, and n-acetylcystein liquid coated zinc oxide quantum dots can significantly reduce the methane gas production by _________
(a) 80 per cent (b) 70 per cent
(c) 40 per cent (d) 90 per cent

131. Nanoparticles might be inhibiting the __________ that produce the greenhouse gases.
(a) Insects (b) Microorganisms
(c) Spider (d) Arthropod

132. Nano additives can be incorporated into premium commercial diesel fuel, which reduces fuel consumption by up to _________ and reduces carbon dioxide emissions by up to __________.
(a) 12 per cent and 17 per cent (b) 22 per cent and 28 per cent
(c) 9 per cent and 18 per cent (d) 10 per cent and 15 per cent

133. ___________ produces nanomaterials that provide a solution to environmental problems
(a) Carbon-based (b) Metal-based
(c) Green nanotechnology (d) Dendrimers

134. _________ are now used in a wide range of applications, including separating petrochemicals, mimicking DNA, and removing heavy metals, fluoride anions, hydrogen, and even gold from water.

(a) Nano pores (b) Nano Tubes
(c) Nano Filters (d) Nano Films

135. Micro porous membranes engineered at the Nano scale using __________ can also be used to capture and store carbon at Carbon dioxide sources.
(a) Manganese Silicates (b) Magnesium Silicates
(c) Nesosilicates (d) Sorosilicates

136. Concentration of atmospheric carbon dioxide is at its highest levels and it can be turn it into __________
(a) Nano pores (b) Carbon nanotubes
(c) Nano Films (d) Nano Filtters

137. Nano systems mediated absorption is the most mature __________, due to its high efficiency and lower cost.
(a) CO_2 separation process (b) O_2 separation process
(c) NO_2 separation process (d) SO_2 separation process

138. __________ is useful for biodiesel production.
(a) Titanium dioxide (b) Ethanol
(c) Jatropha (d) Sweet potato

139. Enzyme immobilization using nanomaterials is another promising application in the __________.
(a) Waste water Industry (b) Biofuel industry
(c) Nano Industry (d) Food Industry

140. Magnetic and silica nanoparticles are widely employed for the immobilization of cellulose enzyme for __________.
(a) Bio pesticide production (b) Bio Fertilizer production
(c) Bioethanol production (d) Bio plastic production

141. Nano materials are used as supports for immobilizing __________
(a) Lipases (b) Fat
(c) Grease (d) Oil

142. Metallic nanoparticles with size range of __________ facilitate the mass transfer coefficient at gas liquid interface, improving CO_2 concentration, thereby influencing algal photosynthesis and accumulation of various valuable bio products.
(a) 5-10 nm (b) 10-100 nm
(c) 1-1000 nm (d) 5-100 nm

143. Application of Zero-valent iron nanoparticles (nZVI) enhances the ______ accumulation in green algae and cyanobacteria
(a) Lipid (b) Fat
(c) Grease (d) Oil

144. Organic Nano clay results in conversion of lipids to biodiesel more than __________
(a) 99 per cent (b) 94 per cent
(c) 98 per cent (d) 90 per cent

145. _______________ in the nanomaterial provides enormous surface area and enhances the diffusion rate of lipase at active sites of reaction.
(a) Mobilization of lipase (b) Immobilization of lipase
(c) Active compound (d) Clay

146. Solid nanomaterial in transesterification process offers high quality ________
(a) Ethanol (b) Methanol
(c) Butanol (d) Biodiesel

147. _________ supported zeolites facilitates hydrogenolysis of triacyl glycerols.
(a) Silver (b) Gold
(c) Nickel (d) Arsenic

148. Biodiesel conversion efficiency of about 91 per cent can be achieved using ___________ during catalytic trans esterification
(a) Calcium oxide nanoparticles (b) Magnesium oxide nanoparticles
(c) Zinc oxide nanoparticles (d) Silver oxide nanoparticles

149. Organic Nano systems are usually _______________ and ecofriendly and they are used commonly in medicinal application.
(a) Degradable (b) Bio degradable
(c) Natural products (d) All of the above

150. Inorganic NPs are generally termed as _______________
(a) Engineered Nanoparticles (b) Degradable Nanoparticles
(c) Core Nanoparticle (d) All of the above

151. Toxic nature of engineered Nanoparticles are attributed to formation of ______ induced damage.
(a) ZVI (b) ROS
(c) NSHA (d) IOMNP

152. ___________________ decides the bactericidal action of engineered Nanoparticles.
(a) Aerobic species (b) Anaerobic species
(c) Reactive oxygen species (d) Both A and B

153. ROS are produced either by ________________.
(a) UV illumination (b) Illumination
(c) Bacterial cells (d) Both A and B

154. Contact of nanoparticles with the cell wall may contribute to the _________ on microorganisms.
(a) Toxicity (b) Non Toxicity
(c) Growth Promoter (d) Neutral

155. ___________ known for its anti-microbial properties
(a) Silver Nanoparticles (b) Copper Nanoparticles
(c) Zinc Nanoparticles (d) Gold Nanoparticles

156. __________ are the next most extensively used ones for its antimicrobial activity.

(a) Silver Nanoparticles (b) Copper Nanoparticles
(c) Zinc Nanoparticle (d) Gold Nanoparticle

157. __________ being the most promising remediation for the contaminated soils.
(a) Zero-valent iron particles (b) Iron particles
(c) Metal Nanoparticles (d) Iron Oxide Nanoparticles

158. _______ are the noble nanoparticle that has gained a lot of attention due to their potential applications in diverse areas from electronics to biomedicine.
(a) Silver (b) Gold
(c) Zinc (d) Iron

159. NPs typically exhibit toxicity to ______________ in terrestrial ecosystem. The metal oxide NPs potentially dissolve in aqueous media, which results in toxic metal ions released into their surrounding media Metal and metal oxide NPs exhibit detrimental effects on the microbial count, metabolic activities and diversity even at very low concentrations.
(a) Bacteria (b) Fungi
(c) Actinomycetes (d) Both A and B

160. Iron Oxide Metal Nanoparticles (IOMNP) potentially stimulate growth in some bacteria and it could also alter the structure of soil bacterial community
(a) Iron Oxide Metal Nanoparticles
(b) Zero-valent iron particles
(c) Reactive oxygen species
(d) Calcium oxide nanoparticles

161. __________ Nanoparticles used in commercial products such as toothpastes and sunscreens.
(a) Zinc oxide (b) Zinc dioxide
(c) Titanium dioxide (d) Silver oxide

162. Toxicity of TiO_2-NPs are associated with the production of Reactive Oxygen Species (ROS) causing an __ in pure cultures of bacteria
(a) Oxidative stress (b) Osmotic stress
(c) Physical damage (d) Both A and B

163. Buckminster (C60) fullerenes were discovered in _________
(a) 1980 (b) 1978
(c) 1985 (d) 1990

164. Increase in concentration of fullerene (up to 500 mg/kg) reduced the amount of _____
(a) Planctomyces (b) Nitrospira
(c) Lysobacter (d) All of the above

165. _____ is based on the leakage of substances/based on measuring the cytoplasmic enzyme
(a) Lactate dehydrogenase (b) Lactose dehydrogenase
(c) Amylase (d) Hydrogenase

166. _______ is performed to assess the geno toxicity effects of NPs on microbes by the action of mobility of nano particles into the cell/nuclear membrane resulting in irreversible DNA damage.
(a) Bio pesticide Assay (b) COMET Assay
(c) Assay (d) Nano Assay

167. Datura metel L. synthesized _______ reduce the larval population of malaria vectors and have no adverse effect on behavior of dragonfly naiads.
(a) Zn NPs (b) Ag NPs
(c) Fe NPs (d) Mn NPs

168. Nano materials decrease the _______ and increase _______ emissions.
(a) Nitrogen and Sulfur (b) Carbon and Nitrogen
(c) Nitrogen and Carbon (d) Carbon and Sulfur

169. NPs synthesized from ________ increased the predator potential of young instars of dragonflya immaculifrons
(a) Datura metel (b) Iron metal
(c) Milkweeds (d) Mucuna

170. ___________ based Nano emulsion was safer to egg parasitoid, T. chilonis than macro-emulsion.
(a) Pungam oil (b) Castor oil
(c) Sesame oil (d) Neem oil

171. Nano materials are effective ___________ against insects and pests.
(a) Nano-cides (b) ZVI
(c) Fungicides (d) Nano spray

172. Algae _________ Affects photosynthetic activity
(a) ZnO (b) Bio pesticides
(c) Ag (d) TiO_2

173. Microbes ___________ Cell disruption.
(a) ZVI (b) CNT
(c) Se (d) Au

174. Plants CNT _________
(a) Cell disruption (b) Inhibited growth
(c) Inflammation (d) Programmed cell death

175. Fish _______ Hyper-accumulation
(a) Au (b) Se
(c) Zno (d) ZVI

176. In Humans model system CNT is
(a) Mesothelioma (b) Inflammation
(c) Death (d) Cell injury

177. ____________is the study of assessing the interaction of nanoparticles either at cellular or organ level to understand its ability to cause damage to the specific organs or even to lethality.

(a) Toxic kinetics (b) Bio kinetics
(c) Toxic dynamics (d) Both A and B

178. Nanoparticles of size more than ____ can't pass through the dermal and nanoparticles of size between _____ passes through damaged skin alone
(a) 45 nm and 18-45 nm (b) 45 nm and 20-45 nm
(c) 20 nm and 18-45 nm (d) 30 nm and 20-50 nm

179. __________ sized particles pass through both intact and damaged cells particles below ________ internalise through the intact cells.
(a) 4 – 10 nm and 4 nm (b) 4 – 40 nm and 7 nm
(c) 4 – 20 nm and 4 nm (d) 2 – 20 nm and 2 nm

180. Per cent of the nanoparticles inhaled (<100 nm) may be deposited in the _____
(a) Lungs (b) Respiratory tract
(c) Eye Balls (d) Stomach

181. The nanoparticles once inhaled along with the air, first interact with the host defence system mainly the __________
(a) Intracellular Route (b) Inter Cellular Route
(c) Pulmonary System (d) Respiratory Lining Fluid

182. Nanoparticle of size __________ can easily reach the alveolar sac.
(a) 2 nm - 0.1 μm (b) 1 nm – 0.01 μm
(c) 1nm – 0.1μm (d) 1 nm – 0.21 μm

183. The most common routes of elimination of nanoparticles from the human system are
(a) Expired air (b) Feces
(c) Urine (d) All of the above

184. Reason for Nano toxicity in most of the living organisms is the excessive generation of ________ leading to oxidative stress as a result of nanoparticle interaction at the sub cellular level.
(a) Electron Hole Pair Theory (b) Active Electronic Stat
(c) Reactive oxygen species (d) Particle Dissolution

185. Most semiconductor chips and transistors are created with __________
(a) Silicon (b) Carbon
(c) Gold (d) Platinum

186. Sumio Iijima discovered __________
(a) Quantum dots (b) Carbon nanotubes
(c) Diode (d) Fullerene

187. Quantum dots are ________
(a) Semiconductor nanocrystals (b) Nanotubes
(c) Carbon sheets (d) Polymers

188. An aggregate of surfactant molecules dispersed in a liquid colloid is called as _____
(a) Micelle (b) Fullerene
(c) Quantum dots (d) Diode

189. Stronger than steel and as flexible as plastic is applicable to ____________
    (a) Carbon nanotubes (b) Fullerenes
    (c) Zeolite (d) Gold
190. Biosynthesis of nanoparticles is a kind of ____________
    (a) Bottom up approach (b) Top down approach
    (c) Mechanical method (d) Chemical method
191. Ball Milling is a __________
    (a) Mechanical method of nano particles synthesis
    (b) Chemical method of nano particles synthesis
    (c) Biological method of nano particles synthesis
    (d) None
192. Particles are considered to be nanoparticles if one of their dimensions is less than __________
    (a) 100 nanometers (b) 1000 nanometers
    (c) 1 nanometers (d) 100 micro meter
193. In microbial method of nanoparticle synthesis, metal compounds are converted into their respective nanoparticles by _____________
    (a) Hydrogenation process (b) Reduction process
    (c) Fixation process (d) Volatilization process
194. There is plenty of room at the Bottom- This famous statement was given by_____
    (a) Alexander Flemming (b) Richard Feynman
    (c) SumioIijima (d) Thomas Muller
195. Kratschmer and Huffmann developed the methodology for the synthesis of ______
    (a) Nano crystals (b) Fullerenes
    (c) Diode (d) Quantum dots
196. Silicon can be converted into a conductor by the process of__________
    (a) Doping (b) Ball milling
    (c) Epitaxy (d) Chemical vapor deposition
197. P or As is added to silicon in_____________
    (a) Nanolithography (b) N type of doping
    (c) Nano composites (d) Nano polymers
198. Molecular beam epitaxy is an advanced form of ______________
    (a) Thermal evaporation (b) Laser evaporation
    (c) Sputtering (d) Chemical vapour deposition
199. Fullerene is composed of______
    (a) Hydrogen atoms (b) Carbon atoms
    (c) Hydrogen and carbon (d) Nitrogen
200. Who first used the term nanotechnology and when?
    (a) Richard Feynman, 1959 (b) Norio Taniguchi, 1974
    (c) Eric Drexler, 1986 (d) Sumio Iijima, 1991

201. What is a buckyball?
(a) A carbon molecule (C60)
(b) Nickname for Mercedes-Benz's futuristic concept car (C111)
(c) Plastic explosives nanoparticle (C4)
(d) Concrete nanoparticle with a compressive strength of 20 nanonewtons (C20)

202. Richard Feynman is often credited with predicting the potential of nanotechnology. What was the title of his famous speech given on December 29, 1959?
(a) There is a tiny room at the bottom
(b) Things get nanoscopic at the bottom
(c) Bottom? What bottom?
(d) There is plenty of room at the bottom

203. In 1996, the Nobel prize in Chemistry is awarded for the discover of ______
(a) Carbon Nano Tubes (b) Fullerene
(c) Quantum dots (d) Silver nano particle

204. Fullerene contains
(a) 20 Pentagons and 12 Hexagons
(b) 12 Pentagons and 20 Hexagons
(c) 22 Pentagons and 20 Hexagons
(d) 10 Pentagons and 20 Hexagons

205. Diamond and graphite are the allotropes of
(a) Carbon (b) Selenium
(c) Oxygen (d) Phosphorous

206. ______________ is an electrical device that converts one form of energy into another in the bio nanosensor
(a) Transformer (b) Transister
(c) Transducers (d) Semiconductor

207. A new carbon molecule with 60 carbon atom is called as ________
(a) Carbon Nanotube (b) Carbon coreshell
(c) Basket ball (d) Fullerene

208. A Microscope which is used to read and write at Nano scale level
(a) Transmission Electron Microscope
(b) Atomic Force Microscope
(c) Scanning Tunnelling Microscope
(d) Confocal Microscope

209. The color of gold nanoparticle __________
(a) Yellow (b) Red
(c) Green (d) White

210. Bio degradable 'Green Plastics' made from ___________
(a) Green polymer (b) Plant material
(c) Synthetic Nanoparticle (d) Nano ploymer

211. Electron Microscope uses ___________ source for making images.
(a) Light (b) Electron
(c) Sound (d) Neutron

212. Semiconductor nanoparticle contains _______________ materials
(a) Conductor and insulator (b) Conductor and Semiconductor
(c) Conductor and conductor (d) Semiconductor and Insulator

213. How many oxygen atoms lined up in a row would fit in a one nanometer space?
(a) None (b) One
(c) Seven (d) Seventy

214. AFM is _______________
(a) Atomic Force Microscopy
(b) Atomic Field Microscopy
(c) Atomic Lateral Force Microscopy
(d) Atomic Frequency Microscopy

215. Nanoparticles are particles whose largest dimension is less than or equal ______
(a) 100 nm (b) 1 nm
(c) 10 nm (d) 1000 nm

216. _________ is a microorganism used to synthesize silver nanoparticle
(a) Lactobacillus sp. (b) Bacillus sp.
(c) E. coli (d) Actinobacter sp

217. Carbon Nanotube was discovered by
(a) Richard E.Smalley (b) Sumio Iijima
(c) One lakh (d) Robert F. Curl Jr.

218. The "Molecular Machinery" was introduced by
(a) Robert F. Curl Jr. (b) K. Eric Drexler
(c) Sir Harold W. Kroto (d) Richard E. Smalley

219. Diode is a
(a) Tube light (b) Semiconductor
(c) Iron Oxide (d) Carbon Nano tube

220. Which of the following is an allotropy of carbon?
(a) Copper (b) Diamond
(c) Gold (d) Quartz

221. Which of the following statement is not true?
(a) Buckminster invented Bucky ball
(b) Sputtering is a physical method
(c) Diode is a semiconductor
(d) Alloys is a mixture of metal and non-metal

222. The most common Buckminster fullerene consists of _______________ atoms
(a) 50 carbon (b) 60 carbon
(c) 70 carbon (d) 80 carbon

223. The Lycurgus cup coated with
(a) Copper nanoparticle (b) Gold nano-particle
(c) Carbon nanotube (d) Gallium nanoparticle

224. Most of the amphiphilic molecules are ____________
(a) Dopants (b) Surfactants
(c) Lyphophilic (d) Thermophilic

225. Reverse micelle is formed when ________________
(a) Water in oil (b) Oil in water
(c) Random mixing (d) None of the above

226. Which of the following material is NOT a natural polymer?
(a) Chitin (b) Silk
(c) Rubber (d) PVC

227. Inert Gas Condensation Technique (IGCT) is a
(a) Mechanical method (b) Chemical method
(c) Physical method (d) Thermal method

228. Identify inert gas is
(a) CO_2 (b) O_2
(c) Ar (d) CO

229. Doping of silicon is done to make a ______________
(a) Conductor to non-conductor (b) Conductor to super conductor
(c) Non-conductor to Conductor (d) Non-conductor to Semi-Conductor

230. Who is Buckminsterfullerene?
(a) Inventor of Bucky ball (b) Architect
(c) Geologist (d) Inventor of CNT

231. Nano robots (nanobots) are
(a) Used in Medicine (b) Used in Agriculture
(c) Used in Environment (d) Used in Energy sector

232. Which one of the statements is NOT true?
(a) Gold at the nanoscale is red
(b) Silicon at the nano-scale is an insulator
(c) Copper at the nanoscale is transparent
(d) Aluminium at the nanoscale is highly combustible

233. Element that is used for "N Type" doping?
(a) Si (b) As
(c) N (d) C

234. How many hydrogen atoms lined up in a row would fit in a one Nano meter space?
(a) 100 (b) 10
(c) One (d) 1000

235. Biological synthesis of Nano -particles can be achieved using
(a) Plants (b) Viruses
(c) Microbes (d) All of the above

236. __________ is used to do the high resolution imaging of an organelle in a plant cell
(a) TEM (b) SEM
(c) AFM (d) Confocal

237. Volatile profile of a cardamom can be precisely determined using __________
(a) GC (b) GC-MS
(c) HPLC (d) GPT

238. Natural nano-particles arise abiotically from earth processes contributing to rock and soil formation, including volcanism and__________
(a) Dissolution (b) Erosion
(c) Precipitation (d) All of the above

239. Famous Indian Scientist who promoted nanotechnology to the public __________
(a) Dr. M.S. Swaminathan (b) Dr. N. Rao
(c) Dr. Annadurai (d) Dr. AP.J. Abdul Kalam

240. A __________is a material that has an electrical resistivity between that of a conductor and an insulator.
(a) Quantum Dots (b) Nano Wire
(c) Nano tube (d) Semiconductor

241. An external electrical field changes a semiconductor's __________
(a) Resistivity (b) Magnetism
(c) Conduction (d) Thermal resistance

242. Molecular beam epitaxy is a __________ preparation of nanoparticles
(a) Physical method (b) Biological method
(c) Chemical method (d) Mechanical method

243. Which of the following is not used for characterization of nanoparticles?
(a) AFM (b) Gas Permeability Tester
(c) TEM (d) Microwave oven

244. SEM is used to study the
(a) Surface property (b) Topography
(c) Internal property (d) All of the above

245. Sputter coater is used for making the sample more
(a) Conductive (b) Non-conductive
(c) Refractive (d) Transparent

246. Nanoparticle synthesis using fungi is a __________
(a) Chemical method (b) Physical method
(c) Biological method (d) Mechanical method

247. Ag NPs possess __________property
(a) Anti-viral (b) Anti-bacterial
(c) Anti-fungal (d) Anti-microbial

248. Top down approach means
(a) Arranging atom by atom (b) Size reduction
(c) Use of chemical reactions (d) Building blocks

249. Sol gel method of synthesis of nanoparticle is a ________________
(a) Chemical method (b) Physical method
(c) Biological method (d) Mechanical method

250. PLGA is a
(a) Biodegradable polymer (b) Naturally occurring resin
(c) Alloy (d) Surfactant

251. Ultra-thin sections in the microtome can be obtained by using
(a) Glass knife (b) Diamond knife
(c) Iron knife (d) All the three

252. Quantum Dots are ______________ dimensional
(a) Two (b) Zero
(c) Three (d) Multi

253. The emission of light from the Quantum Dot depends on
(a) Wave length of the incident light
(b) Size of the quantum dots
(c) Colour of the quantum dot
(d) Composition of the quantum dot

254. Raman spectroscopy works based on the principle of
(a) Molecular vibrations
(b) Type of light source
(c) Atomic force between atoms of molecules
(d) None of the above

255. How many Nano-particles can be made from a virus?
(a) 100 (b) 10
(c) 1000 (d) None of the above

256. ______________ received Nobel Prize in the year 2010
(a) CNT (b) Fullerenes
(c) Quantum dots (d) Graphene

257. All ____________ compound can be detected using GC-MS
(a) Thermolabile (b) Thermolabile & thermostable
(c) Thermostable (d) None

258. Permeability of nano-films can be studied using ____________
(a) GPPT (b) GC-MS
(c) XRD (d) GPT

259. ZnO is having
(a) Anti-oxidant property (b) Anti-fungal property
(c) Anti-bacterial property (d) All of the above

260. Chemical Synthesis of NPs is a ____________ process
(a) Oxidation (b) Reduction
(c) Hydration (d) Addition

261. Microwave synthesis is a __________ method
(a) Mechanical (b) Chemical
(c) Physical (d) Biological

262. Aloe vera gel is known to synthesis ___________ NPs
(a) Au (b) Ag
(c) Ti (d) ZnO

263. NPs can be biologically synthesized using ___________
(a) Plants (b) Bacteria
(c) Fungi (d) All

264. Quantum Dots are used for ________________
(a) Photo-voltic cells (b) Biological label
(c) Drug Delivery (d) All

265. Gold NPs can be synthesized using ________ plants
(a) Rice (b) Lucerne
(c) Tomato (d) Tridax

266. Amul Milk powder is by ___________ technique
(a) Ball Milling (b) Spray drying
(c) Electro spinning (d) Sol-gel

267. Milk is an example for ________ type of colloid
(a) Liquid (b) Emulsion
(c) Solution (d) Emulsified concentrate

268. PLGA is a ________________
(a) Biodegradable (b) FDA approved
(c) Smart delivery vehicle (d) All

269. First State Agricultural University in India to start a dedicated Department for Nano Science & Technology is ________________
(a) PAU (b) GB Pant Nagar
(c) HAU (d) TNAU

270. Carbon Nano-tube is believed to improve seed germination as a result of _________
(a) Seed coat permeation (b) Breaking dormancy
(c) Water absorption (d) a & c

271. Nano-fertilizers are synthesized using ________________natural mineral
(a) Silicates (b) Ferro silicates
(c) Zeolites (d) Bio char

272. Never wash clothes are fortified with ________________ nano-particles
(a) Au (b) Ti
(c) Fe (d) Ag

273. One of the well-recognized nano-materials to promote seed germination is ________
(a) ZnO (b) AgO
(c) CuO (d) TiO_2

274. Institute of Nano Science & Technology in India is located in ________________

(a) New Delhi (b) Hyderabad

(c) Bangalore (d) Mohali

275. ________________ is the major investor for R & D activities in nanotechnology

(a) India (b) Japan

(c) USA (d) Korea

276. Nanotechnology is highly exploited in ____________research

(a) Cancer (b) Urology

(c) Cardiology (d) Dental

277. ________________technique is used to protect the bio inoculants used in agriculture

(a) Spray drying (b) Encapsulation

(c) Electro spinning (d) All

278. In a surfactant molecule, the tail has ____________________ property

(a) Hydrophobic (b) Hydrophilic

(c) Amphiphilic (d) Ionic

279. Nano-thickness of specimens can be achieved by ________________

(a) Sharp Knife (b) Microtome

(c) Ultra microtome (d) Glass Knife

280. Nano-meter refers to one ____________________of a meter

(a) Millionth (b) Thousandth

(c) Billionth (d) Hundredth

281. Inventor of TEM____________________

(a) Ernst Ruska (b) Ernst Ruska and Max Knoll

(c) Max Knoll (d) None

282. ______________ is used to examine the high-resolution imaging of leaf surface

(a) AFM (b) SEM

(c) TEM (d) Confocal

283. Widely used nano-paricle in food system is ______________

(a) Ag (b) CNT

(c) Au (d) Ti

284. The diameter of a Bucky ball is about

(a) 1.1 nm (b) 10 Å

(c) 100 Å (d) 1000 Å

285. The thermal conductivity of a standard SWNT along its length is ____ watt/(m.k).

(a) 3500 (b) 385

(c) 35000 (d) 35

286. Unique approach using electrostatic forces to produce nano-fibre.

(a) Electrophoresis (b) Electro spinning

(c) Electron Gun (d) Electromagnets

287. According to the definition by CRN, nanotechnology is ________
(a) Mechanical engineering (b) Atomic engineering
(c) Newtonian mechanics (d) micro-electronics

288. Nanoparticles prepared by using biologically processed metal ores
(a) Homeopathic medicines (b) Modern antibiotics
(c) Ayurvedic 'Bhasmas' (d) Modern cosmetics

289. Who was the first scientist to describe that substances having nano dimensions possess altogether different and unique properties?
(a) Richard Feynman (b) Eric Drexler
(c) Archimedes (d) Michael Faraday

290. Which of the following does not apply to nanotechnology?
(a) It is a general-purpose technology.
(b) It can be called Green technology.
(c) It involves rearrangement of atoms.
(d) Newtonian mechanics can describe it.

291. The two important properties of Nano substances are ______
(a) Pressure and friction (b) Sticking and friction
(c) Sticking and temperature (d) Temperature and friction

292. When the electric field applied reaches a critical value, the repulsive electrical forces overcome the
(a) Surface tension forces (b) Vander Walls Force
(c) Ionic bonds (d) Covalent bonds

293. The thermal stability of a nanotube is seen up to which K in air __________
(a) 100 (b) 1000
(c) 2000 (d) 3100

294. Nanoparticles of which substance were found on the surface of the sword of Tipu Sultan?
(a) Gold (b) Lead
(c) Carbon (d) Silicon

295. Nano particles of which atom are used to control collateral damage due to explosion?
(a) Copper (b) Aluminium
(c) Carbon (d) Lead

296. The limitations on double bond locations lead to poor delocalization of electrons, increasing the
(a) Molecules reactivity (b) Electron's reactivity
(c) Atoms reactivity (d) Quantum reactivity

297. The magnetic anisotropy of a material can be modelled as uniaxial in character and represented by:
(a) $K = KV \sin^2 \theta$ (b) $K = KV \cos^2 \theta$
(c) $K = KV \tan^2 \theta$ (d) $K = KV \cot^2 \theta$

298. Conductivity of nanoparticles is believed to occur due to:
1. Tunnelling of electrons through the metal core.
2. Hopping of the electrons along the atoms constituting the chain of the legend molecule encapsulating the nanoparticle.

(a) 1 (b) 2
(c) Both (d) None

299. With the help of_____ Robert F. Curl and others discovered fullerene.
(a) Electron microscope (b) Magnetic resonance
(c) Condensation technique (d) Mass spectrograph

300. The smallest cluster of carbon atoms in Bucky balls known till today consists of ____ carbon atoms.
(a) 75 (b) 60
(c) 20 (d) 15

301. The tensile strength of an MWNT is ____ Pa
(a) 63×10^6 (b) 63×10^7
(c) 63×10^8 (d) 63×10^9

302. The electrical conductivity of a nanotube is ____ times that of copper.
(a) 10 (b) 100
(c) 1000 (d) 1/100

303. An MWNT possesses electrical superconductivity up to temperature of ________
(a) 12 K (b) 12°C
(c) 100 K (d) 100°C

304. The wavelength of visible light is ____ nm.
(a) 40-70 (b) 400-700
(c) 4000-7000 (d) 40000-70000

305. The wave length of the photon emission from Quantum Dot depends on the _____
(a) Size of the quantum dots (b) Volume of quantum dots
(c) Radius of quantum dots (d) Number of quantum dots

306. Nano science can be studied with the help of ___________
(a) Quantum mechanics (b) Newtonian mechanics
(c) macro-dynamics (d) Geophysics

307. Greeks and Romans had used nanoparticles in the manufacture of __________
(a) Cosmetics For Eyes (b) Medicines
(c) Metal Articles (d) Hair-Dye

308. ______ are the extensions of Bucky balls.
(a) Geodesic domes (b) Hexagons
(c) Carbon nanotubes (d) AFM and STM

309. Atoms or molecules on a solid surface possess fewer nearest neighbours or coordination numbers, and thus have ___________ exposed to the surface.
(a) Covalent (b) Dangling or unsatisfied bonds
(c) Ionic (d) Vander wall forces

310. In size quantization effect totally an ___________ that occurs in metal and semiconductor nanoparticles.
(a) Electronic effect (b) Magnetic effect
(c) Ionic effect (d) Capillary effect

311. Metals have a cubic or hexagonal close-packed structure consisting of one central atom, which is surrounded in the first shell by 12 atoms, in the second shell by
(a) 12 atoms (b) 22 atoms
(c) 32 atoms (d) 42 atoms

312. The largest cluster of carbon atoms in Bucky balls known till today consists of ____ carbon atoms.
(a) 60 (b) 772
(c) 680 (d) 540

313. The quantum confinement of a carrier in a strip of width w leads to the discretization of energy levels given by
(a) $\epsilon_n = n^2h^2/(8m^*w^2)$. (b) $\epsilon_n = h^2n^2/(8m^*w^2)$.
(c) $\epsilon_n = h^2n^2/(8w^*m^2)$. (d) $\epsilon_n = h^2h^2/(8m^*w^2)$.

314. Nano crystalline particles represent a state of matter in the transition region between ___
(a) Bulk solid and single molecule
(b) Bulk liquid and single molecule
(c) Bulk gas and single molecule
(d) Bulk vapour and single molecule

315. The _______ treats grain boundaries as barriers to dislocation motion, and thus dislocations pile up against the boundary.
(a) Hall-Petch model (b) Moore's Law
(c) Kepler constant (d) Newton 2 law

316. Researchers have found that ________ derivates of fullerenes inhibit the HIV-1 protease
(a) Solvent soluble (b) Fat soluble
(c) Water soluble (d) Lipid soluble

317. In discrete energy levels there exists a _________ and therefore energy absorption is required for an electron to get excited into the conduction energy levels
(a) Forbidden gap (b) Forbidden place
(c) Forbidden atom (d) Forbidden CNT

318. The spot that could hold one transistor 15 years ago can hold today
(a) 1,000 transistors (b) 100
(c) 10 (d) Same number of transistors

319. Moore's second law, predicts
(a) Semiconductor
(b) IC
(c) diode
(d) cost of building a chip manufacturing plant

320. The year in which Gordon Moore foresaw that the rate of progress would reach saturation is
(a) 2015 (b) 2001
(c) 1998 (d) 2003

321. Nanomaterial's can be classified based on dimension
(a) < 100 m Nano rods, nanowires etc
(b) < 100 nm Nanowires , thin films, fibers , platelets etc
(c) < 100 mm Particles, quantum dots, hollow spheres etc.
(d) zero dimensional < 100 mm- bucky ball, carbon nanotube

322. A __________ matériel has a signifiant fraction of crystal grains in the nanoscale.
(a) Nano-crystalline (b) Nano émulsion
(c) Carbon nanotubes (d) Fullerènes

323. The fullerenes are a class of allotropes of Carbon which conceptually are ____ sheets rolled in to tubes or spheres.
(a) Graphene (b) Copper
(c) Platinum (d) Silver

324. Bottom up methods involve the assembly of atoms or molecules into nanostructured arrays. In these methods the raw material sources can be in the form of
(a) Gases (b) liquids
(c) Solids (d) all

325. The popular processes include electric arc-discharge, laser ablation, chemical vapor deposition (CVD), and ball milling-annealing are
(a) Mechanical methods (b) Chemical methods
(c) Biological methods (d) enzymatic methods

326. Metal used in an inert environment is
(a) Gold (b) Silver
(c) Argon (d) Mercury

327. Limitations in ball milling include
(a) Some becomes absorbed into the sample
(b) Effect of contamination tends to be more serious if sample/milling media are similar materials
(c) Using milling media that are of the same material as the sample is the solution if contamination is not critical
(d) But compensation to retain balance of alloy may not be required

328. In fullerenes, numbers of pentagonal rings are necessary and sufficient to affect the cage closure.
(a) 12 (b) 14
(c) 17 (d) 45

329. Quantum dots have radius of about
(a) 1-3 nm (b) 100-200 nm
(c) 1-3 mm (d) 100-200 mm

330. Genetically engineered __________ allow preparation of quantum dot bio composite structures
(a) M13 bacteriophage (b) Bacteria
(c) Streptomyces PVL45 (d) Fungus

331. __________ quantum dots are also used in inorganic/organic hybrid solar cells.
(a) Colloidal (b) Solid
(c) Liquid (d) gas

332. Owing to the material's exceptional strength and stiffness, nanotubes have been constructed with length-to-diameter ratio of up to
(a) 132,000,000:1 (b) 132,000,00:1
(c) 132,000,000:1 (d) 132,000:1

333. What is the size of MEMS?
(a) 10^{-6} to 10^{-4} m (b) 10^{-9} to 10^{-11} m
(c) 10^{-9} to 10^{-11} m (d) 10^{-2} to 10^{-4} m

334. What is the size of a Nano shell?
(a) 100 nm (b) 10 nm
(c) 100 cm (d) 10 cm

335. Who photographed nanotubes for the first time?
(a) Sumio lijima (b) K. Dexer
(c) Feyman (d) Newton

336. Which of the following is an advantage of using high pressure freezing?
(a) It can minimize artifacts prompted
(b) It can produce ice-crystal damage
(c) samples without preparation can be used
(d) Better preservationof antibody for immune-hydro chemical studies can be achieved

337. A higher vacuum of __________ or higher in the electron gun in high- resolution or field-emission TEMs
(a) 10^{-4} to 10^{-7} Pa (b) 10^{-6} to 10^{-11} Pa
(c) 10^{-3} to 10^{-9} Pa (d) 10^{-1} to 10^{-14} Pa

338. When using ultra microtome for sectioning, what is the purpose of using glass knife?
(a) To hold the sample blocks.
(b) To section the specimens into thin slice.
(c) It allows us to monitor the process of sectioning.
(d) To trim the cutting edge of sample blocks so as to make it flat.

339. Why should we add water to the diamond cutting boat before sectioning?
(a) To remove dust from the boat
(b) To sharpen the diamond knife
(c) To facilitate the collection of sample slices
(d) To prevent dehydration of sample after sectioning

340. Which of the following is used as a fixative in TEM?
    (a) Sodium cacodylate (b) PBS
    (c) Glutaraldehyde (d) Uranyl acetate
341. Transmission electron microscope uses a beam of _______ to create an image of the specimen.
    (a) Light (b) Protons
    (c) Sound waves (d) Electrons
342. The microscope gives the detailed 3-dimensional architecture of cell surfaces.
    (a) Transmission electron microscope
    (b) Compound microscope
    (c) Scanning electron microscope
    (d) Phase contrast microscope.
343. Sodium cacodylate is used for TEM sample preparation as a:
    (a) Fixative (b) Stain
    (c) Buffer (d) Fixative and stain
344. For viewing which of the following would a TEM not be a good choice?
    (a) Antigen (b) Actin filament
    (c) Ribosomes (d) 3D external surface of mitochondria
345. Approximately _________ of the atoms in ___________ are found on the surface
    (a) 100 %, Bulk materials (b) 50 %, Nanostructures
    (c) Both a and b (d) None of the above
346. Generally, for semiconductors such as ZnO, CdS, and Si; when the bandgaps lie in the _________ spectrum, changing bandgap with _______ means a change in _________.
    (a) Infrared, charge, colour
    (b) Visible, size, colour
    (c) Ultraviolet, temperature, colour
    (d) None of the above
347. The surface to volume ratio for a cube in Nano scale with a side 'a' is
    (a) a/6 (b) 6a
    (c) 6/a (d) 12a
348. Nano science is dominated by two electron phenomena
    (a) Quantum Size effect and Tunnelling
    (b) Collision and Velocity
    (c) Both a and b
    (d) None of the above
349. Langmuir films, LB films, and SAMs are all examples of ____________ and ____________ nanostructures
    (a) Inorganic thin films, 3D (b) Organic thin films, 2D
    (c) Both a and b (d) None of the above

350. ____________ based processing techniques fail at the ________ because one cannot succeed in reducing features in the images produced to ________ than the ____________ of the light used
(a) Heat, Nanoscale, more, temperature
(b) Velocity, Nanoscale, less, wavelength
(c) Pressure, less, wavelength
(d) Light image, Nanoscale, less, wavelength

351. Sol-gel method is used for the fabrication of
(a) Metal oxides (b) Carbon nanotubes
(c) Quantum dots (d) Bucky ball

352. Complete the equation Si(OR.4 + $H_2O \rightarrow$ HO–Si(OR.3 + ________
(a) R–OH (b) R-COOH
(c) R-CH_2OH (d) R-CHO

353. In processing of ceramic nanomaterials uncontrolled flocculation of powders due to
(a) Ionic bond (b) Covalent bond
(c) Vander Walls Force (d) None of the above

354. Drying the gel by means of low temperature treatments (25-100 °C), it is possible to obtain porous solid matrices called
(a) Phytogels (b) Hydrogels
(c) Aerogels (d) Xerogels

355. Particle Size Analyzers are based on technologies, such as
(a) Light scattering of the particles
(b) Analysis of Brownian motion
(c) Both
(d) None

356. Particle size distribution is a property unique to powders and is an important physical property for determination of powders.
(a) Behaviour and nature (b) Taste and smell
(c) Texture and structure (d) Melting and boiling point

357. All particle sizing techniques measure a one-dimensional property of a particle and relate this to the size of an
(a) "Equivalent sphere" (b) "Equivalent square"
(c) "Equivalent cube" (d) "Equivalent size"

358. Photon Correlation Spectroscopy can measure size range of
(a) 1nm to 5μm (b) 1mm to 5mm
(c) 1cm to 5cm (d) 1m to 5m

359. In microwave assisted synthesis of nanoparticles dipolar interactions occur within
(a) Non polar molecules (b) Polar molecules
(c) Both (d) None

360. Heat is generated by microwave is due to
    (a) Molecular collision and friction
    (b) UV rays
    (c) Gamma rays
    (d) Tungsten filament
361. Electromagnetic waves generated by microwave is due to
    (a) 300 MHZ to 300 GHZ (b) 30 MHZ to 30 GHZ
    (c) 3 MHZ to 3 GHZ (d) 300000 MHZ to 300000 GHZ
362. Advantages of microwave based heating include
    (a) Vessel wall transparent to microwave
    (b) Instant on/off
    (c) Direct in core heating
    (d) All of the above
363. Anti-reflection coatings with Nano particles reduce
    (a) Reflection losses (b) Light Transmission
    (c) Brightness (d) Refractive Index
364. Silicon nano tips on solar panel cells, which are nanostructures mimicking
    (a) Moth's Eye (b) Lizard's eye
    (c) Spider's Eye (d) None of These
365. Most widely for making antireflective
    (a) Nano zinc oxide (b) Silver Nitrate
    (c) Nano Titanium dioxide (d) None of these
366. Solar panel conversion efficiency
    (a) 10% (b) 20%
    (c) 30% (d) 40%
367. Self-cleaning of solar panels is based on
    (a) Oil Effect (b) Lotus Effect
    (c) Angle effect (d) None of these
368. Water contact angle to super hydrophobic resulting in self-cleaning of solar panels
    (a) >150° (b) <150°
    (c) 150° (d) None of these
369. Photo Conversion Efficiency: Single Cell
    (a) 27 % (b) 28 %
    (c) 29 % (d) None of these
370. Photo Conversion Efficiency: Multi Cell
    (a) 43% (b) 44 %
    (c) 45 % (d) None of these
371. DSSC
    (a) Dye Sensitized Solar Cells (b) Dye Sterilized Solar Cells
    (c) Dye Systematic Solar Cells (d) None of these

372. Layer lubrication which reduce friction and wear in turbines.
(a) Titanium dioxide (b) Zinc oxide nanowires
(c) Molybdenum disulphide (d) None of these

373. Nano crystals with crystallite size of 20 nm gave 99 per cent conversion of soybean oil to biodiesel
(a) Iron oxide (b) Calcium oxide
(c) Zinc oxide (d) None of these

374. Offer large surface area, small size, and large number of active adsorption sites, compared to the conventional adsorbents
(a) Nano-catalysts (b) Nano membranes
(c) Nano-adsorbents (d) None of these

375. Degrade organic contaminants
(a) Nano-catalysts (b) Nano membranes
(c) Nano-adsorbents (d) None of these

376. Used for water treatment can be carbon based, polymeric, magnetic, or non magnetic based, and metal oxide-based including modified compounds such as oxide composites and Zeolites.
(a) Nano-catalysts (b) Nano membranes
(c) Nano-adsorbents (d) None of these

377. Replaced reverse osmosis due to the lower operational costs and reduced energy consumption and have high rejection for most multivalent ions and organic molecules with low molecular weight (Nano zeolites, nano-Ag, nano-TiO , and 2 carbon nanotubes (CNTs).
(a) Nano-catalysts (b) Nano membranes
(c) Nano-adsorbents (d) None of these

378. Non-toxic, High adsorption capacity, Selectivity to low concentration of pollutants, Cost Effective and Reusable
(a) Nano-catalysts (b) Nano membranes
(c) Nano-adsorbents (d) None of these

379. NPs comprise of elemental iron or other metals in combination with a metal catalyst, such as platinum, gold, nickel, and palladium to increase the reduction rate.
(a) Metallic NPs (b) Bimetallic NPs
(c) Tandem NPs (d) None of these

380. Commercially available Bimetallic NPs
(a) Silver and iron (b) Palladium and iron
(c) Titanium and iron (d) None of these

381. Highly structured polymer molecules consisting of a central core from which branches originate.
(a) Dendrimers (b) Monomers
(c) Polymers (d) None of these

382. A research environment that supports advanced data acquisition, data storage, data management, data integration, data mining, data visualization and other com-

puting and information processing services over the Internet

(a) Environmental sensing (b) Nano sensing
(c) Electro sensing (d) None of these

383. "Bhasma" is in the size of ___________

(a) 1-10 nm (b) 5-50 nm
(c) 10-100 nm (d) None of these

384. NDDS

(a) Novel drug delivery system (b) Nano drug delivery system
(c) Neuro drug delivery system (d) None of these

385. Application of nanotechnology in medicine, draws on the natural scale of biological phenomena to produce precise solutions for disease prevention, diagnosis, and treatment.

(a) Nano medicine (b) Nano-diagnosis
(c) Nano-treatment (d) None of these

386. Science and technology of diagnosing, treating and preventing oral and dental diseases, relieving pain, preserving and improving dental health using nanostructured material.

(a) Nano-dentistry (b) Nano-diagnosis
(c) Nano-treatment (d) None of these

387. Nanobots are

(a) 5–100 nm wide (b) 50–100 nm wide
(c) 5–10 nm wide (d) None of these

388. Science of engineered Nano devices and nanostructures that deals with their effects in living organisms.

(a) Nano-mycology (b) Nano-diagnosis
(c) Nano-toxicology (d) None of these

389. NPs of Ag and ZnO are

(a) Harmful (b) Toxic
(c) Very toxic (d) Extremely toxic

390. C60 fullerenes and CuO NP are

(a) Harmful (b) Toxic
(c) Very toxic (d) Extremely toxic

391. SWCNTs and MWCNTs are

(a) Harmful (b) Toxic
(c) Very toxic (d) Extremely toxic

392. TiO_2 NP are

(a) Harmful (b) Toxic
(c) Very toxic (d) Extremely toxic

393. TiO_2 Algae Model system

(a) Affected photosynthetic activity
(b) Inhibited growth

(c) Cell disruption
(d) None of these

394. CeO_2 Algae Model system
(a) Affected photosynthetic activity
(b) Inhibited growth
(c) Cell disruption
(d) None of these

395. ZVI Microbes Model system
(a) Affected photosynthetic activity
(b) Inhibited growth
(c) Cell disruption
(d) None of these

396. CuO Microbes Model system
(a) Affected photosynthetic activity
(b) Inhibited growth
(c) Cell disruption
(d) Antibacterial activity

397. Ag Microbes Model system
(a) Affected photosynthetic activity
(b) Inhibited growth
(c) Cell disruption
(d) Increased cell permeability

398. ZnO Microbes Model system
(a) Increased cell permeability (b) Inhibited growth
(c) Cell disruption (d) 100% mortality

399. CNT Plants Model system
(a) 100% mortality (b) Inhibited growth
(c) Cell disruption (d) Programmed cell death

400. ZnO Plants Model system makes
(a) Programmed cell death (b) Inhibited growth
(c) Cell disruption (d) Inhibited seed germination

401. Se Fish Model system promotes
(a) Inhibited seed germination (b) Inhibited growth
(c) Cell disruption (d) Hyper-accumulation

402. ZVI Fish Model system promotes
(a) Hyper-accumulation (b) Inhibited growth
(c) Cell disruption (d) Morphologic changes

403. Al_2O_3 Rodents Model system promotes
(a) Morphologic changes (b) Inhibited growth
(c) Cell disruption (d) Inflammation

404. TiO_2 Rodents Model system promotes
(a) Inhibited growth (b) Inter growth
(c) Cell disruption (d) Inflammation

405. ZVI Rodents Model system makes
(a) Inflammation (b) Inhibited growth
(c) Cell disruption (d) Toxic to nerve cells

406. CNT Rodents Model system produce
(a) Toxic to nerve cell (b) Inhibited growth
(c) Cell disruption (d) Necrosis

407. Au Rodents Model system
(a) Necrosis (b) Inhibited growth
(c) Cell disruption (d) Moved from mother's placenta to fetus

408. Magnetite Humans Model system
(a) Neurodegenerative diseases (b) Inhibited growth
(c) Cell disruption (d) Moved from mother's placenta to fetus

409. Au Humans Model system
(a) Neurodegenerative diseases
(b) Cell disruption
(c) Moved from mother's placenta to fetus
(d) Penetrated into the sperm head and tails causing fragmentation

410. Which of these historical works of art contain nanotechnology?
(a) Lycurgus cup
(b) Damascus steel swords
(c) Medieval stained-glass windows in churches
(d) All of the above

411. Neem leaves is used for synthesis of __________ Nano particles
(a) Silver (b) Gold
(c) Copper (d) Zinc

412. Banana Peel is used to synthesis __________ Nano particles
(a) Copper (b) Silver
(c) Cadmium sulphide (d) Gold

413. Green tea is used to synthesis __________ Nano particles
(a) Silver (b) Gold
(c) Copper (d) Zinc

414. During activation phase, the metal ions reduced to zero oxidation state and self-organization of metal atoms started through
(a) Chemical process (b) Nucleation process
(c) Physical process (d) Reduction process

415. Defects are usually absent in either metallic or ceramic clusters of nanoparticles because dislocations are basically

(a) Unstable (b) Mobile
(c) Both a & b (d) more

416. ________ is a single crystal within a bulk/thin films form
(a) Grain (b) Creep
(c) Crystal (d) Plasticity

417. __________ is defined as the property of material remain deformed after the force is removed
(a) Grain (b) Creep
(c) Crystal (d) Plasticity

418. ___________ is the tendency of a solid material to move slowly or deform permanently under the influence of mechanical stresses.
(a) Flow (b) Creep
(c) Fall (d) Plasticity

419. __________ contain internal interfacial defects, known as grain boundaries, where the lattice orientation changes
(a) Grain (b) Creep
(c) Crystal (d) Plasticity

420. __________ gives a quantitative description of an increase in the yield stress of a polycrystalline material as its grain size decreases.
(a) Planks law (b) Hall-Petch law
(c) de-Broglie's law (d) Lamberts law

421. __________ is defined as the energy portion of a thermodynamic system available to do work
(a) Gibbs free energy (b) Raman's law
(c) Eienstein mass energy (d) Planks energy concept

422. Ferromagnetism disappears and transfers to super paramagnetic in the nanometer scale due to their huge.
(a) Low surface area (b) Huge surface area
(c) Temperature (d) Pressure

423. Electrical properties gradually changes with _________ particle size.
(a) Increasing (b) Decreasing
(c) Remains same (d) constant

424. ___________are allotropes of carbon.
(a) CNT (b) Nano wire
(c) Quantum dot (d) Nano film

425. When a nanoparticle is much ___________ than the wave length of light, coherent oscillation of the conduction band electrons induced by interaction with an electromagnetic fiel(d)
(a) Bigger (b) Smaller
(c) Very bigger (d) Quicker

426. When a nanoparticle is much smaller than the wave length of light, coherent oscillation of the conduction band electrons induced by interaction with an electro-

magnetic field. This resonance is called ___________

(a) SSR (b) SRS
(c) SPR (d) RPS

427. The changes gold–blue–purple–red are largely geometric ones that can be explained with ______________

(a) Eisenstein's theory (b) Planks theory
(c) Gauss theory (d) Mie theory

428. The changes gold–blue–purple–red are largely geometric ones that can be explained with Mie theory which describes light ___________ by a sphere.

(a) Scattering (b) Diffraction
(c) Reflection (d) Transformation

429. _____________ microorganism involved in the synthesis of silver Nano particles.

(a) Fusarium oxysporium (b) Actinobacter sp.
(c) Verticillium sp (d) Thermophilic bacteria

430. ______________ microorganism involved in the synthesis of cadmium nano particles

(a) Fusarium oxysporium (b) Actinobacter sp.
(c) Verticillium sp (d) Thermophilic bacteria

431. ___________ microorganism involved in the synthesis of gold nano particles

(a) Fusarium oxysporium (b) Actinobacter sp.
(c) Verticillium sp (d) Lactobacillus sp.

432. How many electrons are revolving around the silicon in outer orbit?

(a) Four (b) One
(c) Seven (d) Six

433. Which among the following statement is incorrect?

(a) Photography is an early example of nanotechnology
(b) MIE Theory was proposed by German physicist
(c) Fieldelectron microscope was invented by Ervin Mueller
(d) Dr. James found the double helix structure of RNA

434. Which of these consumer products is already being made using nanotechnology methods?

(a) Fishing lure (b) Golf ball
(c) Sunscreen lotion (d) All of the above

435. If you (1.7 m) were to shrink yourself down until you were only a nanometre tall, how thick would a sheet of paper appear to you?

(a) 170 meters (b) 1.7 kilometres (a bit more than a mile)
(c) 17 kilometres (d) 170 kilometres

436. What is graphene?

(a) A new material made from carbon nanotubes
(b) A one-atom thick sheet of carbon
(c) Thin film made from fullerenes
(d) A software tool to measure and graphically represent nanoparticles

437. Which of these well-known phrases from Star Trek depends on the (fictional) use of nanotechnology?
   (a) Beam me up, Scotty.
   (b) Tea Earl Grey. Hot.
   (c) You will be assimilated. Resistance is futile.
   (d) All of the above

438. What is grey goo?
   (a) A hypothetical substance composed of out-of-control self-replicating nanobots that consumes all living matter on Earth
   (b) The feeder material used to grow grey nanoparticles in the laboratory
   (c) Toxic by-product resulting from the synthesis of carbon nanotubes
   (d) Waste product from the production of nanoglue made from the membranes on the feet of the Madagascan Grey Gecko

439. Which one of these condiments is unique due to the Nano scale interactions between its ingredients?
   (a) Ketchup (b) Mustard
   (c) Mayonnaise (d) All of the above

440. Presents of Nano robots (nanobots) contemporarily ___________
   (a) Do not exist yet
   (b) Exist in experimental form in laboratories
   (c) Are already used in Nano medicine to remove plaque from the walls of arteries
   (d) Will be used by NASA in the next unmanned mission to Mars

441. What has been the cumulative budget for the U.S. National Nanotechnology Initiative since its inception in 2001?
   (a) \$187 million (b) \$1.9 billion
   (c) \$14 billion (d) \$29 billion

442. Plasmonics is___________
   (a) A field of Nano photonics that holds the promise of molecular-size optical device technology
   (b) The science of fluorescent nanoparticles used in modern fireworks
   (c) A hypothetical science used in science fiction weaponry (plasma cannons)
   (d) The technology used to design and build the laser-guided photonic gyroscopes used in aviation.

443. Optical tweezers ___________
   (a) Are used to remove facial hair with miniaturized laser beams
   (b) Use light to manipulate particles as small as a single atom
   (c) Are a nanotechnology-based tool for stamp collectors
   (d) Don't exist

444. A silver coin with a diameter of 4 cm (such as the U.S. silver dollar) contains 26.96 grams of coin silver and has a surface area of about 27.7 square cm. If the same 26.96 grams of coin silver were divided into particles 1 nanometer in diameter, what would their combined surface area be?

(a) 11.4 square meters (b) 140 square meters
(c) 1,400 square meters (d) 11,400 square meters

445. What exactly is a quantum dot?
(a) A fictional term used in science fiction for the endpoints of wormholes
(b) The sharpest possible tip of an Atomic Force Microscope
(c) A semiconductor nanostructure that confines the motion of conduction band electrons, valence band holes, or excitons in all three spatial directions.
(d) Unexplained spots that appear in electron microscopy images of nanostructures smaller than 1 nanometer

446. Nano materials are the materials with at least one dimension measuring less than ________
(a) 1 nm (b) 10 nm
(c) 100 nm (d) 1000 nm

447. A material with one dimension in Nano range and the other two dimensions are large is called ________
(a) Quantum wire (b) Micro-material
(c) Quantum well (d) Quantum dot

448. The melting point of particles in nano form ________
(a) Increases (b) Remains same
(c) Decreases (d) Increases then decreases

449. The initial tools used to help launch the Nano science revolution were ________
(a) Binoculars (b) Scanning probe instruments
(c) Microscope (d) Interferometer

450. Which of the processes of materials was not described as Nanotechnology?
(a) Separation (b) Creation
(c) Processing (d) Consolidation

451. The size of atoms is nearly ________
(a) 0.01 nm (b) 0.1 nm
(c) 1 nm (d) 10 nm

452. Which form of carbon is man-made?
(a) Graphite (b) Microtubules
(c) Fullerene (d) Diamond

453. "There is a plenty of room at the bottom." This was stated by
(a) Eric Drexler (b) Richard Feynman
(c) Albert Einstein (d) Issac Newton

454. N-type semi-conductor is obtained by doping Si with
(a) B (b) P
(c) Ga (d) Sb

455. A nanometer is
(a) A unit of measure equal to one billionth of a meter
(b) A unit of measure equal to one millionth of a meter

(c) A unit of measure equal to one billionth of a millimetre
(d) A unit of measure equal to one meter

456. ____________ is the study of fundamental principles of molecules and structures with at least one dimension roughly between 1 and 100 nanometers.
(a) Femtotechnology (b) Astrophysics
(c) Nano science (d) Pico technology

457. Moore's first law states that
(a) Amount of space required to install a transistor shrinks roughly by 18 times
(b) Cost of building a chip manufacturing plant doubles with every other chip generation
(c) Cost of building a chip manufacturing plant triples with every other chip generation.
(d) Amount of space required to install a transistor expands roughly by 18 times

458. Which among the following statement is incorrect?
(a) Richard Feynman is a chemist
(b) Eric Drexler wrote the book 'Engines of Creation'
(c) Moore's law surpassed in 1994
(d) The word dendra means seed

459. Which among the following statement is correct?
(a) The Nobel prize for chemistry was awarded in 2001
(b) Quantum dots are engineered nanoparticles
(c) Electronic devices exploiting the spin of electrons, in addition to charge is TEM
(d) Cost of building a chip manufacturing plant triples with every other chip generation

460. The prefix nano means?
(a) One billionth (b) One crore
(c) One millionth (d) Two billionth

461. Which year carbon nano tube is discovered?
(a) 1992 (b) 1991
(d) 1994 (c) 1993

462. The populations of transistors used per square inch in the IC's will double every 18 months is called ____________
(a) Circuit (b) Moore's Law
(d) Semi-conductor (d) Quantum dots

463. The simplest possible semiconductor device is ____________
(a) Transistor (b) Diode
(c) Quantum (d) CNT

464. A quantum dot is a _________
(a) Nanotube (b) Semiconductor
(c) Nano wire (d) Nano cluster

465. Hydrophilic is
(a) Water liking (b) Ether liking
(c) Lipid liking (d) Oil liking

466. Moore's law states that the increase in complexity of microprocessors and other types of semiconductor integrated circuits will double every_________
(a) 16 months (b) 36 months
(c) 10 months (d) 28 months

467. Which among the following statement is incorrect?
(a) Indians were using nanotechnology from 18th century
(b) In 1996 Noble prize for the discovery of Bucky ball
(c) Nano particles turns Aluminium in to highly explosive
(d) Quantum dots are engineered nano particles

468. Which among the following statement is correct?
(a) Zeolite is an organic porous material
(b) Nanotubes behave like non metal
(c) Carbon nanotube was developed by IBM
(d) Chip shrinks by roughly half every 36 months

469. Molecular nano technology was proposed by
(a) Eric Drexler (b) Moore
(c) Feynman (d) Aviram

470. Who prepared and explained nanotubes for the first time_____________
(a) SumioTijima (b) Richard Smalley
(c) Eric Drexler (d) Richard Feynman

471. Who was the first scientist to describe those substances having Nano dimensions possess altogether different and unique properties?
(a) Richard Feynman (b) Eric Drexler
(c) Archimedes (d) Newton

472. 1 nm _____________
(a) 1000 picometres (b) 1/1,000,000,000
(c) 10 angstroms (d) All of the above

473. The world's most famous Lycurgus cup explicit the _______ Property of nanomaterial.
(a) Mechanical (b) Optical
(c) Electrical (d) All of the above

474. The Lycurgus cup was created by___________ at 4th Century
(a) Babylonians (b) Greek
(c) Romans (d) Egyptians

475. The Lycurgus cup was made up of __________ glass
(a) Tempered glass (b) Dichroic glass
(c) Silicon glass (d) None

476. Fullerenes are a molecular form of pure carbon discovered
(a) 1985 (b) 1987
(c) 1959 (d) 1928

477. Molecular nanotechnology, originated in the mind was proposed by _______
(a) Eric Drexler (b) Sutton
(c) Feynman (d) Darwin

478. Angstrom (Å): Unit of length used to measure atoms and molecules __________
(a) 0.1 nanometer (nm) (b) 0.4 nanometer (nm)
(c) 0.5 nanometer (nm) (d) 0.9 nanometer (nm)

479. The wave length of infra-red radiation is ____________
(a) 400-700 nm (b) 750 nm- 1500 nm
(c) 10nm – 400 nm (d) 0.1 cm – 100 cm

480. The size of nanoparticles is between _____ nm.
(a) 0.1 to 10 (b) 1 to 100
(c) 0.01 to 1 (d) 100 to 1000

481. Identify the correct statement
(a) Non-metal semiconductor is Si
(b) Non-metal semiconductor is silver
(c) Non-metal semi-conductor is gold
(d) Non-metal semiconductor is Cd Se

482. Example for metallic nano materials
(a) Hydrogen (b) Oxygen
(c) Carbon (d) Silver

483. Identify the semiconductor Nano particle from the following
(a) Cd se (b) Au
(c) Mg (d) Ag

484. Quantum dots are example for
(a) 0 dimensional (b) 1 dimensional
(c) 3 dimensional (d) 2 dimensional

485. The chip performance would double every
(a) 10 months (b) 18 months
(c) 12 months (d) 20 months

486. Structure of tiny substances which can only be seen under electron microscopes is called?
(a) Nano carbon (b) Nano Technology
(c) Nano scale (d) Nano structure

487. Who is discovered STM &TEM
(a) Zurich , Gerd Binnig , Heinrich rogrer
(b) Allen &Allen, omelianky, flemming, zurich
(c) Browen, starkey, conn, frank
(d) Eric, Taniguchi, Feynman ,seden

488. Nanometers in an inch
   (a) 25,400,000 (b) 26,400,000
   (c) 27,400,000 (d) 28,400,000
489. Founder of the Intel corporation
   (a) Gordon (b) Browen
   (c) Zurich (d) Eric
490. Atoms, molecules, complexes are attached to the exterior of the cage?
   (a) Endohedral fullerenes (b) Exohedral fullerenes
   (c) Endo and Exo fullerens (c) Endo and Exo fullerens
491. The dimension of nanoparticle is
   (a) < 100 nm (b) > 100 nm
   (c) = 100 nm (d) 1 - 100 nm
492. Smallest molecule is
   (a) O_2 (b) H_2
   (c) N_2 (d) CO_2
493. Spherical fullerenes are also called
   (a) Spherical ball (b) Bucky ball
   (c) Cricket ball (d) Ball
494. Consist of two or more layers of carbon
   (a) Multi wall nano tube (b) Single wall nano tube
   (c) Multi & single wall nano tube (d) Fullerene
495. Carbon nanotubes are
   (a) Armchair (b) Quantum dots
   (c) Conductor (d) Bucky ball
496. Which one is cheapest method for synthesis of Nano particle
   (a) Physical method (b) Chemical method
   (c) Biological method (d) Mechnical method
497. What kind of material used in biological method
   (a) Silicon (b) Carbon
   (c) Microorganisms (d) Microorganism and plant extract
498. Etching out of crystal plants from bulk materials to Nano materials
   (a) Bottom-up approach (b) Top-down approach
   (c) Top up approach (d) Bottom down approach
499. Chemical and biological methods which type of approach
   (a) Bottom-up approach (b) Top up approach
   (c) Bottom down approach (d) Top down approach
500. ________ ideas and concepts behind Nano science and nanotechnology started with a talk entitled
   (a) There's of room at the bottom
   (b) There is plenty of room at the top

(c) The bottom of room at the bottom
(d) There is plenty of room at the bottom

501. __________ abbreviations of STM & AFM
(a) Scanning tunnelling microscope & Atomic force microscope
(b) Scanning turning microscope & atomic focus microscope
(c) Scanning tunnelling microscope & atomic flame microscope
(d) Source tunnelling microscope & atomic force microscope

502. ______________ is discovered Moore's law
(a) Gordon (b) Brown
(c) Walia (d) Reiter

503. ______________ depends on semi-conductor
(a) Radio waves (b) UV waves
(c) gamma rays (d) X rays

504. ______________ is the heart of electronic device
(a) Cd (b) Si
(c) Se (d) Ag

505. ______________ type doping of silicon is a good conductor?
(a) N Type (b) P type
(c) N &P type (d) A & B type

506. __________ is the world biggest silver producing country
(a) Mexico (b) India
(c) USA (d) Sri lanka

507. ____________ isolated from silver mines has been shown to produce silver nano particles
(a) Pseudomonas aeruginosa (b) Pseudomonas stutzeri
(c) Bacillus subtlis (d) Pseudomurein palladium

508. Alkalo thermophilic actinomycete synthesized are ____________ nano particles
(a) Gold (b) Silver
(c) Al (d) Zi

509. __________had better metal accumulating ability and late assimilating it as Nano particles
(a) *Medicago sativa* (b) *Brassica juncea*
(c) *Helianthus annus* (d) *Capsicum annum*

510. *Azadirachta indica* are synthesised ____________
(a) Extracellular and intra cellular (b) intracellular
(c) Extra cellular (d) Periplasmic

511. A variety of super hard Nano composites can be made of ______________
(a) Nitrogen, boron, carbon (b) Nitrides, Borides, Carbides
(c) Nitrides, borides, minerals (d) Borides, plasma, minerals

512. __________ have been found to have excellent mechanical properties
(a) Bucky ball (b) Diode
(c) Carbon nano tube (d) Quantum dots

513. __________ is a fundamental property of all atoms
(a) Diamagnetism (b) Ferro magnetism
(c) Para Magnetism (d) Ferri magnetism

514. The sol gel process is a ____________ technique
(a) Wet - chemical (b) Dry- chemical
(c) Wet & dry chemical (d) wet technique

515. Both single wall nanotube and multi walled Nano tube have been produce _______
(a) Chemical method (b) Mechanical method
(c) Optical method (d) Electrical method

516. Nano tube structure is armchir then the electrical properties are ___________
(a) Semi conducting (b) Metallic
(c) copper (d) Silicon

517. ______________ is the property of those materials which are intrinsically magnetically ordered and which develop spontaneous magnetization without the need to apply.
(a) Ferromagnetism (b) Magnetism
(c) Para magnetism (d) Diamagnetism

518. The light microscope was developed from the galilean telescope during the century
(a) 16th century (b) 17th century
(c) 18th century (d) 19th century

519. _____________ ring bonds can be considered double bond in fullerene
(a) 6 : 5 (b) 6 : 4
(c) 6 : 2 (d) 6: 6

520. Which one is not related for Max knoll ____________
(a) Electrical engineer (b) Electron microscope
(c) German (d) 1932

521. The wavelength of electron is ____________
(a) 0.124 A0 (b) 0.125A0
(c) 0.126 A0 (d) 0.127 A0

522. Which of the particle size measurement techniques are correct or incorrect
(a) Dynamic light scattering (b) Photon correlation spectroscopy
(c) Elastic light scattering (d) Co precipitation

523. The most of magnetic properties of a material can be derived from is ________
(a) Hysteresis loop (b) Hydrolysis
(c) Hydroxyl (d) Hydrophyte

524. The magnetic properties of nano particles are ____________
(a) Diamagnetism (b) Paramagnetism
(c) Ferromagnetism (d) Non ferromagnetism

525. It is possible to synthesize downstream processing procedure is /are _________
(a) Metallic and oxide nano particles (b) Gold
(c) Silver (d) Gold and silver

526. The chemical synthesis of nanoparticles is / are ____________
(a) Micro emulsions (b) Sono chemical synthesis
(c) Electro spinning (d) Ball milling

527. The carbon nanotubes have been classified as ____________
(a) Armchair (b) Zig - Zag
(c) Chiral (d) Bucky ball

528. Structural family of carbon nanotube is /are ____________
(a) Fullerene (b) Semi-conductor
(c) Quantum dots (d) Diode

529. How much larger fullerenes have been found on carbon atoms ____________
(a) 180 (b) 190
(c) 240 (d) 550

530. Which one of the following is wrong for Nano materials ____________
(a)Amorphous (b) Crystalline
(c) Polymeric (d) Solvothermal

531. Which type of ball milling used for laboratory ____________
(a) Industrial ball milling (b) Planetary ball milling
(c) Horizontal ball milling (d) Vibrating ball milling

532. Which one is correct shape of nano materials ____________
(a) Spheres (b) Circle
(c) Triangle (d) Rectangle

533. The investigation of light matter interaction at the nano scale is called ________
(a) Nano optics (b) Nano assembler
(c) Nano probe (d) Nano machine

534. Which Nano scale device used to understand the properties of materials and systems
(a) Nano optics (b) Nano assembler
(c) Nano probe (d) Nano machine

535. A molecular machine made to perform mechanical functions is called
(a)Nano optics (b) Nano assembler
(c) Nano probe (d) Nano machine

536. In which year the first molecular electronic device was patented by Aviram and Seiden of IBM.
(a)1994 (b) 1974
(c) 1986 (d) 1980

537. Scientists of IBM succeed in moving and precisely positioning individual molecules at room temperature in ____________.
(a) January 1976 (b) January 1986
(c) January 1996 (d) January 1974

538. According to SI Unit description 1/100 of a meter that is around an inch is called
(a) millimetre (b) centimetre
(c) micrometre (d) Nanometer

539. In SI Unit one meter is approximately equal to
(a) Three feet or one yard (b) 1/1000 of a meter
(c) Around half an inch (d) 1/1,000,000 of a meter

540. In nano scale the prefix "tera " has the origin from
(a) Micros (b) Monster
(c) Megas (d) Gigas

541. The power term 10^3 in nano scale has prefix name as
(a) Milli (b) mega
(c) kilo (d) giga

542. Anti-adhesive/ anti stain coatings, applied films , virus applied films having dimension as
(a) All three dimension < 100 nm (b) 0 dimensions < 100 nm
(c) 1 dimension < 100 nm (d) 2 dimensions < 100 nm

543. The example for 1 dimensional Nano particle is
(a) Thin films (b) Carbon nanotubes
(c) Solid crystalline bulk particles (d) QD

544. Which of the following is good example for non-metal Nano material
(a) Au and Ag (b) H_2
(c) Carbon (d) Cdse

545. Which one has Two - terminal electronic component that conducts electric current in only one direction is
(a) Transistor (b) Solar cells
(c) C_{60} (d) Diode

546. Which amplifies and switch electronic signals is called
(a) Diode (b) Transistors
(c) C60 (d) Solar cell

547. In a pure __________ crystal is nearly an insulator and very little electricity will flow thought it
(a) Transistor (b) Diode
(c) Silicon (d) C_{60}

548. Which one of the following is not a P- type doping of silicon
(a) B or Ga is added to the silicon (b) They have three outer electrons
(c) They have five outer electrons (d) Holes can conduct current

549. Who invented QDs in colloidal solutions in 1980
(a) Alexei Ekimov (b) Louis Brus
(c) Robert Brown (d) Dalton

550. Zeta potential influences rapid coagulation or flocculation in the range of
(a) zero to ± 5 (b) ± 10 to ± 30
(c) ± 30 to ± 40 (d) ± 40 to ± 60

551. Zeta potential influence's incipient in stability in the range of
(a) zero to ± 5 (b) ± 10 to ± 30
(c) ± 30 to ± 40 (d) ± 40 to ± 60

552. Zeta potential influences moderate in stability in the range of
(a) zero to ± 5 (b) ± 10 to ± 30
(c) ± 30 to ± 40 (d) ± 40 to ± 60

553. Zeta potential influences good stability in the range of
(a) zero to ± 5 (b) ± 10 to ± 30
(c) ± 30 to ± 40 (d) ± 40 to ± 60

554. Compound microscope invented by
(a) Eric Drexler (b) Robert hooke
(c) Harold Croto (d) Richard Smalley

555. ____________ is commonly used as a source of electron
(a) Metal tungsten (b) Metal oxide
(c) Metal aluminium (d) Metal chloride

556. Nano tube consist of two layers of carbon called as
(a) MWNT (b) SWNT
(c) Nano film (d) Thin film

557. Nano tube consist of just one layer of carbon called as
(a) MWNT (b) SWNT
(c) Nano film (d) Thin film

558. Nano tubes are made of which type of carbon compound
(a) Carbon (b) Graphene
(c) Carbon Nano film (d) Carbon Thin film

559. Bulk synthesis is very difficult for ________ type of nano tube synthesis
(a) MWNT (b) SWNT
(c) Nano film (d) Thin film

560. Which of the following is such a challenge when designing molecular machinery
(a) Thermal noise (b) Nano design
(c) Quantum dots (d) film

561. As of public record at the end of 2002, which country was making the greatest annual investment in molecular nanotechnology research
(a) Russia (b) United States
(c) Japan (d) South Korea

562. As computing, medicine, space science, and the military. What was the name of this ground-breaking book?
(a) Smaller is Better (b) Engines of Creation
(c) A Crowded Blueprint (d) The Atomic Cookbook

563. _____ contains nano particles prepared by using biologically processed metal ores.
(a) Homeopathic medicines (b) Modern antibiotics
(c) Ayurvedic Bhasmas (d) Modern cosmetics

564. The diameter of human hair is ____ nm.
(a) 50000 (b) 75000
(c) 90000 (d) 100000

565. The cut-off limit of human eye is ____ nm.
(a) 2000 (b) 5000
(c) 10000 (d) 50000

566. The size of E.Coli bacteria is ____ nm.
(a) 2000 (b) 5000
(c) 50 (d) 90

567. The size of RBC is ____ nm.
(a) 50 (b) 90
(c) 2000 (d) 5000

568. Electrons confined in 1 direction
(a) Electrons can easily move 3 dimensions
(b) Electrons can easily move 2 dimensions
(c) Electrons can easily move 1 dimension
(d) Electrons can easily move 0 dimension

569. Electrons confined in 2 directions
(a) Electrons can easily move 3 dimensions
(b) Electrons can easily move 2 dimensions
(c) Electrons can easily move 1 dimension
(d) Electrons can easily move 0 dimension

570. Electrons confined in 3 directions
(a) Electrons can easily move 3 dimensions
(b) Electrons can easily move 2 dimensions
(c) Electrons can easily move 1 dimension
(d) Electrons can easily move 0 dimension

571. Particles that show size quantization effect are sometimes called
(a) Nano materials (b) Silicon
(c) Copper (d) Q particle or quantum dots

572. Material size &lt100 nm having lower
(a) Melting point (b) Bond length
(c) Pressure (d) Lattice formation

573. Solid particles are very small such a decrease in
(a) Surface area (b) boiling point
(c) pressure (d) bond length

574. Energy required to create a unit area of new surface
(a) surface area (b) Free energy
(c) Surface energy (d) Surface pressure

575. Magnetic hard drives are based on a Nano structured device called
(a) Magneto resistance (b) Magneto power
(c) Magneto frequency (d) Magneto resistance sensor

576. CNT can be excellent?
(a) Strong tube (b) Conductors
(c) Semiconductor (d) Metal

577. CNT has good conductivity higher than ____________ times than copper
   (a) 100 (b) 10
   (c) 1000 (d) 1
578. CNT have thermal conductivity as high as
   (a) Copper (b) Diamond
   (c) Silicon (d) Gold
579. CNT are held together by ____________ force forming a rope like structure
   (a) Electric (b) Conductive
   (c) Thermal (d) Vander Walls
580. Poly lactic acid is a strong
   (a) Hydrophobic (b) Hydrophilic
   (c) Amphiphilic (d) Neutral
581. Inorganic core shell Nano particle the core and shell made by using
   (a) Inorganic material (b) Organic material
   (c) Lanthanum material (d) Semiconductor material
582. The process of packaging of Nano scale active ingredient with in tiny envelope or shell
   (a) Core shell packaging (b) Encapsulation
   (c) Rolling (d) Shielding
583. Kind of electron microscope which is used to study internal structure of cells is
   (a) SEM (b) TEM
   (c) Light microscope (d) Compound microscope
584. Scanning electron microscopy (SEM) is best used to study
   (a) Small internal cell structures (b) Surface morphology
   (c) Compound (d) All of the above
585. The most important property of nanomaterial is
   (a) Force (b) Friction
   (c) Pressure (d) Temperature
586. Nanotechnology, in other words, is
   (a) Carbon engineering (b) Atomic engineering
   (c) Small technology (d) Microphysics
587. Basic source of magnetism is ____________
   (a) Charged particles alone (b) Movement of charged particles
   (c) Magnetic dipoles (d) Magnetic domains
588. Example for dia -magnetic materials
   (a) Super conductors (b) Alkali metals
   (c) Transition metals (d) Ferrites
589. Which of the following is an example of top-down approach for the preparation of nanomaterials?
   (a) Gas phase agglomeration (b) Molecular self-assembly
   (c) Mechanical grinding (d) Molecular beam epitaxy

590. The properties like melting point, solubility, colour, etc changes on varying the
(a) Size (b) Composition
(c) Surface properties (d) None of the mentioned

591. Bucky balls are a type of ______________ shaped like ________________
(a) Graphite; flat layers (b) Graphite; a crystal
(c) Fullerene; a geodesic dome (d) Fullerene; a hollow cylinder

592. The diameter of hydrogen atom is ______ nm.
(a) 10 (b) 1
(c) 0.1 (d) 0.01

593. The hardest material found in nature is ______.
(a) Steel (b) Topaz
(c) Diamond (d) Quartz

594. ______ are the extentions of Bucky balls.
(a) Geodesic domes (b) Hexagons
(c) Carbon nanotubes (d) AFM and STM

595. The width of carbon nanotube is ______nm.
(a) 1 (b) 1.3
(c) 1.55 (d) 10

596. Example for para-magnetic materials
(a) Super conductors (b) Alkali metals
(c) Transition metals (d) Ferrites

597. Example for ferro-magnetic materials
(a) Super conductors (b) Alkali metals
(c) Transition metals (d) Ferrites

598. The sol-gel process is a wet-chemical technique (also known as Chemical Solution Deposition) widely used recently in the fields of materials science and ________.
(a) Electronic engineering (b) Metallurgy
(c) Ceramic engineering (d) Tissue engineering

599. Carbon atoms make ____ type of bond with other carbon atoms.
(a) Covalent (b) Ionic
(c) Metallic (d) Hydrogen

600. Which ratio decides the efficiency of Nano substances?
(a) Weight/volume (b) Surface area/volume
(c) Volume/weight (d) Pressure/volume

601. Stable gold Nano particles with molecular protection were made in solution during ______
(a) 1995 (b) 1998
(c) 1994 (d) 2004

602. Moore's Law surpassed in ______________
(a) 2011 (b) 2001
(c) 1991 (d) 2000

603. Noble prize for physics in 2008 was awarded jointly to Albert Fert and Peter Gruenberg for the discovery of ________
(a) Giant Magneto resistance (b) Assembler
(c) Bucky balls (d) Gold nano particles

604. Building larger objects starting from smaller building blocks is called as ________.
(a) Femtotechnology (b) Top down
(c) Bottom up (d) Pico technology

605. Sol gel process is a ____________ that uses a chemical solution or colloidal particles to produce an integrated network
(a) Wet chemical technique (b) Physical technique
(c) Biological technique (d) Dry chemical technique

606. __________ is a spherical cage structure of carbon containing 60 carbon atoms.
(a) Assembler (b) Buckminster fullerene
(c) Dendrimer (d) Quantum dots

607. Manipulating materials on a scale of elementary particles is known as ________
(a) Femto technology (b) Top down
(c) Bottom up (d) Pico technology

608. ________ is a generic term to describe micron scale electrical / mechanical devices.
(a) MIMS (b) Molecular mechanics
(c) MEMS (d) Molecular wire

609. ______________ is a process where particles are ejected from the targeted material due to bombardment of energetic particles.
(a) Ball milling (b) Physical vapor deposition
(c) Sputtering (d) Pulverizing

610. When a mixture of powders (of different metals or alloys/compounds) are milled together, during which material transfer is involved to obtain a homogenous alloy is called
(a) Mechanical Alloying (b) Mechanical milling
(c) Sputtering (d) Pulverizing

611. ____________ is an all-purpose device for guiding chemical reactions by positioning molecules.
(a) Assembler (b) Buckminsterfullerene
(c) Dendrimer (d) Quantum dots

612. Fabrication of nanostructures without biology or wet chemistry
(a) Wet Nanotechnology (b) Dry nanotechnology
(c) Femtotechnology (d) Pico technology

613. Nanonbot is otherwise known as
(a) Nanomachine (b) Nanobubbles
(c) Quantum dots (d) Fullerenes

614. Electronic device exploiting the spin of electrons in addition to charge is known as
(a) Smart materials (b) Spintronics
(c) Nanobubbles (d) Fullerene

615. ________ is an example of naturally occurring nanoparticles
(a) Sea spray (b) Composites
(c) Volcanic (d) All of the above

616. ________ is an example for engineered nanoparticles
(a) Sea spray (b) Composites
(c) Welding fumes (d) Quantum dots

617. Fullerene consists of ________ carbon atoms
(a) 30 (b) 16
(c) 60 (d) 20

618. Fullerene formed from
(a) 20 hexagonal and 12 pentagonal faces
(b) 12 hexagonal and 20 pentagonal faces
(c) 15 hexagonal and 50 pentagonal faces
(d) 12 hexagonal and 21 pentagonal faces

619. Carbon Nanotubes are otherwise known as
(a) Bucky balls (b) Bucky tubes
(c) Nanobubble (d) Fullerene

620. The diameter of MWNT is ________
(a) 5 - 80 nm (b) 1-2 nm
(c) 80 -100 nm (d) 50 - 100 nm

621. Tunnelling current is a ________ property of Nanoparticles
(a) Physical (b) Chemical
(c) Optical (d) Electrical

622. In nanoparticles ferromagnetism disappears and transfers to ________ due to high surface energy
(a) Paramagnetism (b) Superparamagnetism
(c) Ferrimagnetism (d) Dimagnetism

623. In most of the chemical synthesis the nanoparticles have been formed by ____
(a) Sputtering (b) Mechanical alloying
(c) Mechanical milling (d) Reduction and decomposition of precursors

624. ________ are typical precursors in sol – gel method
(a) Titanium (b) Metal chloride
(c) Alkaloid (d) Both C and D

625. Which of the following technique is used for particle size analysis
(a) Laser diffraction analysis (b) Image analysing
(c) Dynamic light scattering (d) All of the above

626. Dynamic Light scattering technique measures the size of particle in the range of ___

(a) 0.3 nm – 8 (b) 0.3 nm – 8 mm

(c) 0.3 μm – 8 μm (d) 3 nm – 8 μm

627. A potential exists between the particle surface and dispersing liquid is ________

(a) surface potential (b) Zeta potential

(c) Scattering (d) Dispersion

628. Which among the following statement is correct?

(a) Gun is the source of electron beam

(b) Condensor lens is to place the sample

(c) Sample chamber is to converge the electron beam

(d) Camera is to direct the image to the screen

629. Which among the following statement is incorrect?

(a) Specimen for TEM must be thick

(b) In SEM sample is scanned line by line

(c) SEM voltage ranges from 60- 300000 volts

(d) In TEM the electron beams are broad and satic

630. Which among the following statement is correct?

(a) Nanometer is one millionth of a mm

(b) Nanometer is one billionth of a mm

(c) Nanometer is one trillionth of a mm

(d) Nanometer is one millionth of a metre

631. Which among the following statement is correct?

(a) Richard Feynman is a chemist

(b) The word dendra means seed

(c) Moore's law surpassed in 1994

(d) Eric Drexler wrote the book 'Engines of Creation'

632. Which among the following statement is incorrect?

(a) The attachment of a chemically active moieties to an inert molecule / entity is called functionalisation

(b) Electronic devices exploiting the spin of electrons, inaddition to charge is SEM

(c) Quantised resistance observed in some semiconductors at low temperatures is called quantum hall effect

(d) A method by which the molecular potential energy function is computed to understand and predict various properties of the system is molecular mechanics

633. Which among the following statement is correct?

(a) Nanomaterial can be occur only naturally

(b) Electronic devices exploiting the spin of electrons, in addition to charge is SEM

(c) All the three arbitrary are above 100 nm in three dimensional nano particles.

(d) The physical properties of Nanoparticles are not related to size and shape

634. Which among the following statement is incorrect?
   (a) Fullerene is a spheroidal molecule consists of 16 carbon atom
   (b) Fullerene is a closed cage structure with a football shape
   (c) Noble prize in Chemistry was awarded for the discovery of fullerene
   (d) Fullerene are capable to fight against HIV virus
635. Which among the following statement is incorrect?
   (a) CNT was discovered by Eric Drexler
   (b) CNT composed of entirely C atoms
   (c) Quantum dots were discovered in 1980
   (d) Quantum dots are engineered nanoparticles
636. Which among the following statement is correct?
   (a) The term QD was given by Ekimovin
   (b) QDs are used in drug discovery
   (c) Fullerene has 12 pentagon and 20 hexagon
   (d) The size of the QD ranges from 50 – 150 nm
637. Which among the following statement is incorrect?
   (a) SEM gives morphology of the image
   (b) SEM works very slow
   (c) Most SEM requires minimum preparation action
   (d) SEM requires special skill to operate.
638. The electron gun produces an electron beam when _____ wire is heated by current?
   (a) Tungsten (b) Copper
   (c) ZnO (d) TiO_2
639. When the beam touches the sample it produces ___________
   (a) X rays (b) Secondary electrons
   (c) Back scattered electron (d) All of the above
640. Back scattered electrons are detected by ___________
   (a) Energy dispersive Spectrometer (b) Solid state detector
   (c) Everhart Thornley detector (d) None of the above
641. X rays are detected by ________
   (a) Energy dispersive Spectrometer
   (b) Solid state detector
   (c) Everhart Thornley detector
   (d) None of the above
642. In CNT's the carbon atoms are arranged in a hexagonal pattern bonded together with
   (a) Vander Walls Force (b) Covalent bonding
   (c) Ionic bonding (d) Hydrogen bonding
643. Carbon Nano tubes have
   (a) High electrical and high thermal properties
   (b) Low electrical and low thermal properties

(c) High electrical and low thermal properties
(d) Low electrical and high thermal properties

644. Which of the following structure have metallic properties
(a) Arm chair (b) Zig zag
(c) Chiral (d) None of the above

645. __________ is defined as the property of material remains deformed after the force is removed
(a) Grain boundaries (b) Creep
(c) Plasticity (d) Hardness

646. When no other molecules are adsorbed onto the Nano crystallites, the surface atoms are ____
(a) Highly unsaturated (b) Highly saturated
(c) Remains unchanged (d) Less unsaturated

647. If the nanotube structure is armchair then the electrical properties are ______
(a) Non-metallic (b) Semi conducting
(c) Metallic (d) Poor conducting

648. When electrons are confined in 3 directions, then electrons can easily move in ______
(a) 0 dimension (b) 1 dimension
(c) 2 dimensions (d) 3 dimension

649. Microwave is a _________ of synthesis of nanomaterial
(a) Physical (b) Chemical
(c) Mechanical (d) Hybrid

650. The another name of lithography is _________
(a) Self-assembly (b) Bottom up
(c) Top down (d) Synthesis

651. Ball Milling is a typical of _________ process
(a) Top down (b) Bottom up
(c) Synthesis (d) Self assembly

652. Nano scale materials are
(a) At least one dimension sized from 1 to 100 nanometers
(b) At least one dimension sized from 1 to 1000 nanometers
(c) At least one dimension sized from 1 to 10 nanometers
(d) At most one dimension sized from 1 to 100 meters

653. Sol-Gel Method is a __________ process of synthesis of nanomaterials
(a) Physical (b) Chemical
(c) Phychochemical (d) Hybrid

654. Quantum mechanics deals with ___________
(a) Behavior of matter (b) Atom
(c) Nucleus (d) All of the above

655. Human Visible wave length region
(a) 200-500nm (b) 300-500nm
(c) 400-700nm (d) 100-300nm

656. Resolving power = ____________
(a) Wavenumber of light used/4 (numerical aperture of objective)
(b) Wavelength of light used/4 (numerical aperture of objective)
(c) Wavelength of light used/2 (numerical aperture of objective)
(d) Wavenumber of light used/2 (numerical aperture of objective)

657. Anti-stain coatings are _________
(a) 0 – dimensional (b) 1- dimensional
(c) 2- dimensional (d) 3-dimensional

658. The De-Broglie wavelength λ= __________
(a) h/m (b) h/p
(c) p/h (d) p/mv

659. 1 nm is __________
(a) 1/109 m (b) 1/108 m
(c) 1/107 m (d) 1/106 m

660. Natural originated Nano particles
(a) Sea spray (b) Volcanic ash
(c) Viruses (d) All

661. Parthenium leaves are used for the synthesis of _________
(a) Silver (b) Iron
(c) Gold (d) Copper

662. Carbon nanotubes discovered in the year _______ by the japanese electron micropist Sumio lijima
(a) 1981 (b) 1991
(c) 2001 (d) 1971

663. The element which is used for doping silicon to make it as a conductor
(a) Sodium (b) Calcium
(c) Phosphorus (d) Magnesium

664. A measure of the charge on a particle surface is called _________
(a) Potential (b) Zeta potential
(c) Surface potential (d) Surface charge

665. TEM is the method used to ________ a nanoparticle
(a) Characterize (b) Synthesis
(c) Intersect (d) Section

666. Atomic force microscopy was invented in the year __________
(a) 1989 (b) 1982
(c) 1972 (d) 1978

667. Electrical conductivity of the silica can be managed by the process called ______
(a) Lopping (b) Electrical circuit
(c) Doping (d) All

668. Mesoscale device can be used in the range of __________
(a) 10-8 m (b) 10-9 m
(c) 10-10m (d) 10-7m

669. Based on origin of the Nano materials are classified in to ______ types
(a) Four (b) Three
(c) Six (d) Five

670. The smaller the quantum dot will be close to which end
(a) Blue (b) Red
(c) Yellow (d) Green

671. Strength of carbon nanotube was determine by the _______ bond.
(a) SP^3 (b) SP^2
(c) SP^4 (d) SP^6

672. In Gibbs free energy $\Delta G = \Delta H - T\Delta S$; S denotes
(a) Enthalpy (b) Entropy
(c) Work (d) Temperature

673. In magnetic properties of nano material, ferromagnetism disappears and transferred in to __________
(a) Super paramagnetism (b) Paramagnetism
(c) Anti ferromagnetism (d) Supermagnetism

674. Most commonly used catalyst in SWCNT synthesis
(a) Au (b) Ag
(c) Fe (d) Mg

675. Luminescence is an emission of ______
(a) Light (b) Sound
(c) Electron (d) Neutron

676. Most of the Nano particle was synthesis by using which method ______
(a) Physical (b) Mechanical
(c) Chemical (d) Biological

677. The diameter of the human DNA is ________
(a) 1nm (b) 0.1nm
(c) 100 nm (d) 2.5nm

678. Sputtering process is a ___________
(a) Physical method (b) Mechanical methods
(c) Chemical method (d) Biological method

679. The bio sensor that measures changes in heat is
(a) Piezoelectric (b) Optical
(c) Calorimetric (d) Electro Chemical

680. The process of changing the conductivity of the material by adding small impurities
(a) Vulcanization (b) Doping
(c) Galvanizing (d) Coating

681. The number of carbon atoms consisted in most common Bucky ball
(a) 80 carbon atoms (b) 70 carbon atoms
(c) 20 carbon atoms (d) 60 carbon atoms

682. Co precipitation method of synthesis of Nano particle
(a) Physical method (b) Chemical method
(c) Mechanical methods (d) Biological method

683. Which transducer converts biological energy into electrical energy?
(a) Semiconductor (b) Fullerenes
(c) CNT (d) Biosensor

684. What is meant by Nano Lithography?
(a) Nano machine (b) Nano scale writing
(c) Molecular manufacturing (d) Nano scale fluids

685. Hall-petch law gives a quantitative description of an __________
(a) Increase in the yield stress of a polycrystalline material as its grain size increases
(b) Decrease in the yield stress of a polycrystalline material as its grain size decreases
(c) Increase in the yield stress of a polycrystalline material as its grain size decreases
(d) Decrease in the yield stress of a polycrystalline material as its grain size increases

686. Which wafers used to grow CNT by chemical vapour deposition?
(a) Titanium (b) Silicon
(c) Gold (d) Gallium

687. Strain is the amount of measure of _________
(a) Formation (b) Deformation
(c) Plasticity (d) Adhesion

688. SPR occurs when the size of metal particle is smaller than light
(a) Transmission (b) Viscosity
(c) Wave length (d) Frequency

689. Giant magneto resistance is the application of ________
(a) Paramagnetic NP (b) Ferromagnetic NP
(c) Super paramagnetic NP (d) Super ferromagnetic NP

690. PVD refers to ____________
(a) Partial vapour deposition (b) Partial vaccum deposition
(c) Physical vapour deposition (d) Physical vaccum deposition

691. Doping strongly affects the ______________ properties of semiconductor Nano crystal
(a) Optical (b) Thermal
(c) Physical (d) Chemical

691. The size of virus is ________
(a) 10 nm (b) 0.1 nm
(c) 1 nm (d) 100 nm

692. PVD is a ______________method of Nano particle synthesis
(a) Physical (b) Chemical
(c) Electrical (d) Optical

694. QDs produce ________ number of electrons for every photon absorbed from the sun
(a) 3 (b) 2
(c) 1 (d) 5

695. Semiconductor chips and transistors are primarily made of ________
(a) Silicon (b) Aluminium
(c) Iron (d) Gallium

696. An example for two dimensional Nano particles is _______
(a) Nano-film (b) Nano-wire
(c) Nano-transistor (d) Nano-semi conductor

697. ________ are different structural forms of the same element
(a) Autotropes (b) Allotropes
(c) Chemotropes (d) Lithotropes

698. Nanomaterial wherein all the dimensions are measured within the 100nm are ____ dimensional
(a) Zero (b) Two
(c) Five (d) Four

699. Single walled CNT differs based on symmetry into ________
(a) Arm chair (b) Zig zag
(c) Helical (d) All of the above

700. _______ are not confined to the Nano material range in any dimension
(a) Bulk (b) 10^{-9}
(c) One billionth of meter (d) 10OA

701. In Bucky ball structure each pentagon is surrounded by __________ hexagons
(a) Five (b) Six
(c) Four (d) Three

702 _________ is commonly used for thin film deposition
(a) PVD (b) Laser ablation
(c) Arc discharge (d) Sputtering

703 _________ wavelength lasers are used in laser ablation technique
(a) Wider (b) Longer
(c) Narrower (d) Shorter

704. In sputter deposition, usually ________ are incident on a target at a high energy.
(a) Ar+ (b) Fe+
(c) K+ (d) Mg+

705. Who developed mechanical milling in 1970 __________
(a) Norio (b) John Benjamin
(c) Richard Feynman (d) Louis

706. Packaging of Nano scale active ingredient within tiny envelope or shell
(a) Quantum dot (b) Bucky ball
(c) Nano coating (d) Encapsulation

707. The shell protection imparts __________ functional properties to the nanomaterial
(a) Solubility (b) Stability
(c) Both a & b (d) None of the above

708. Polymer coated core shell nano particle synthesis includes
(a) Polymerisation & adsorption (b) Polymerisation & sputtering
(c) Sputtering & adsorption (d) Polymerisation & desorption

709. Quantum dots made from semiconductor materials like
(a) Cadmium, lead, zinc (b) Cadmium, nickel, zinc
(c) Cadmium, arsenic, nickel (d) Cadmium, lead, arsenic

710. Size of Quantum dots varies from
(a) 100nm-200nm (b) 200-300nm
(c) 5-50nm (d) 50-100nm

711. Quantum dots synthesised by colloidal method also known as
(a) Chemical synthesis (b) Mechanical method
(c) Physical method (d) Biological method

712. QDs in glass matrix were discovered by
(a) Norio (b) Alexi Ekimov
(c) Richard Feynman (d) Louis

713. A device that blocks current on one direction while letting current flow in another direction __________
(a) Diode (b) Anode
(c) Anode & Diode (d) None of the above

714. Conductivity of silicon can be changed through a process called __________
(a) Diode (b) Anode
(c) Dipping (d) Doping

715. Types of doping are
(a) N-doping & P-doping (b) P-doping
(c) NP-doping (d) N- Doping

716. In N-type doping __________ are the impurities added to silicon
(a) P or As (b) B or Ga
(c) Ni or Zn (d) Cd or Ni

717. In P-type doping __________ are the impurities added to silicon
(a) P or As (b) B or Ga
(c) Ni or Zn (d) Cd or Ni

718. Normal micelle is ____________
(a) Oil in water (b) Water in oil
(c) Water in water (d) Oil in oil

719. ______________ is an example for micelle
(a) Water (b) Detergent
(c) Water and detergent (d) Oil

720. _____________ is an example for micelle in human body
(a) Fat soluble vitamins (b) Minerals
(c) Vitamins (d) Fat soluble minerals

721. A micelle is an aggregate of surfactant molecules dispersed in a ___________
(a) Liquid colloid (b) Solid-gas phase
(c) Gaseous phase (d) Liquid-solid phase

722. Nanotubes are ___________
(a) Spiral fullerenes (b) Cylindrical fullerenes
(c) Helical fullerenes (d) Elliptical fullerenes

723. Nano buds have been obtained by adding ______________ to carbon nanotubes
(a) Fullerenes (b) Bucky ball
(c) Sensor (d) Carbon

724. CNTs are held together ___________ by forming a rope-like structure.
(a) Thermal force (b) Vander Walls Force
(c) Atomic force (d) Magnetic force

725. Typical nanomaterial includes carbon-based fullerenes Eg. Is
(a) Buckminsterfullerene (b) Bucky balls
(c) C60 (d) Semiconductor

726. Ultra high vaccum is ___________
(a) 10^{-3} torr (b) 760 torr
(c) $10^{-3} - 10^{-8}$ torr (d) $10^{-8} - 10^{-12}$ torr

727. The tensile strength of carbon nanotube is ________ times that steel
(a) 10 (b) 25
(c) 100 (d) 6

728. Process of nanoparticle synthesis using fungi is ________
(a) Physical (b) Chemical
(c) Mechanical (d) Biological

729. __________ is a plant source for synthesis of gold nanoparticles
(a) Ocimum sanctum (b) Musa paradisiacal
(c) Camellia sinensis (d) Diospyros kaki

730. The element known as the building block of all molecules of life is _________
(a) Phosphorus (b) Nitrogen
(c) Potassium (d) Carbon

731. The surfactants are mostly _________ in nature
(a) Solid (b) Gel
(c) Liquid (d) Gaseous

732. __________ method is a liquid phase method for the synthesis of nanoparticle
(a) Physical (b) Chemical
(c) Mechanical (d) Biological

733. In Bucky ball each carbon atoms forms covalent bonds with ____ other carbon atoms
(a) Four (b) Three
(c) Two (d) One

734. Multi walled carbon nanotubes are ____________
(a) Less purity (b) High purity
(c) Conductor (d) Insulator

735. Quantized resistance observed in some semiconductor at low temperature is under _________ effect
(a) Quantum hall (b) Field
(c) Quantization (d) Hall pelch

736. An example of scanning probe microscope
(a) AFM (b) TEM
(c) SEM (d) EM

737. Single walled CNT are classified in _______ types based on symmetry properties
(a) Three (b) Six
(c) Eight (d) Ten

738. Armchair structure of CNT has _______ type of character
(a) Non-metallic (b) Metallic
(c) Insulator (d) Conductor

739. In hall petch effect one grain size reaches about _____ grain boundaries start to slide
(a) 11 nm (b) 16 nm
(c) 10 nm (d) 3 nm

740. ____________ is a line defect within a crystal which arise during crystal growth
(a) Dissolution (b) Dislocation
(c) Solution (d) Emulsion

741. Migration of crystal defect is caused by ___________
(a) Adsorption creep (b) Absorption creep
(c) Diffusion creep (d) Dissolution creep

742. Hall petch effct says the smaller the grain size ______ the material
(a) Weeker (b) Linear
(c) Rounder (d) Stronger

743. The quasi continuous density of states in the valance and the conduction bands splits into discrete electronic levels
(a) Field effect (b) Quantum hall effect
(c) Size quantization effect (d) Hall Petch effect

744. Stable metal cluster is a gold particle with 55 atoms first reported by
(a) G. Schmic 1985 (b) G. Schmic 1986
(c) G. Schmic 1981 (d) G. Schmic 1990

745. Metallic nanotubes can carry an electrical current ________ more than a copper
(a) 100 times (b) 1000 times
(c) 10 times (d) 2 times

746. What is the origin of the colour?
(a) Surface plasmons (b) Plasmons
(c) Oscillation plasmons (d) Oscillation

747. Oscillation of the conduction band electron induced by interaction with an electromagnetic field
(a) Electromagnetic resonance (b) Surface Plasmon resonance
(c) Surface resonance (d) Plasmon resonance

748. The silicon Nano crystals in PS that emits visible light vary in size from
(a) 15 -20 Å (b) 6 -8 Å
(c) 10-13 Å (d) 10-15 Å

749. The electron hole pair radiatively recombines to yield light at approximately
(a) 650 nm (b) 750 nm
(c) 850 nm (d) 550 nm

750. PL and EL emission spectra are
(a) Homogeneous (b) Heterogeneous
(c) Inhomogeneous (d) No change

751. Self-assembly and coalescence of atoms leads to crystal plans the formation of nanoparticles
(a) Bottom up (b) Top up
(c) Bottom? What bottom (d) Top? What top

752. Formation of mesoporous TiO_2 by ________ process
(a) Saponification (b) Sonication
(c) Wet gel (d) Ball milling

753. The reduction of particle size by high energy ball milling is termed as a ___ approach
(a) Top up (b) Top down
(c) Bottom up (d) Bottom down

754. Bottom up approaches for synthesis is otherwise called as
(a) Physical & Mechanical (b) Physical & Chemical
(c) Chemical & Biological (d) Biological & Mechanical

755. ZnO Nano particles are synthesized under which temperature
(a) 160°C (b) 170°C
(c) 180°C (d) 135 °C

756. The two basic types of micro emulsions are
(a) Direct & Indirect (b) Direct & Reversed
(c) Reversed & Stable (d) Stable & Unstable

757. Co-precipitation method is the simplest chemical reaction involving inorganic salts and ________
(a) Solute (b) Organic salt
(c) Solvent (d) Solution

758. Ultrasound irradiation is used in which type of Nano material synthesis
(a) Sol-gel (b) Polyol
(c) Electro spinning (d) Sonochemical

759. Microorganisms are digested to collected nanoparticles
(a) Intracellular synthesis (b) Extracellular synthesis
(c) Culturing media (d) Metal synthesis

760. Particles are three dimensional objects, and unless they are perfect sphere
(a) Suspension (b) Emulsion
(c) Solution (d) Solute

761. Larger the particle size
(a) Faster Brownian motion (b) No change in Brownian motion
(c) Slower Brownian motion (d) Less Brownian motion

762. The rate of movement of particles in a liquid medium is related to translational diffusion coefficient
(a) Stoke's law (b) Stoke's Einstein's law
(c) Boltzmann's law (d) Moore's law

763. Stability behaviour of the colloids were good when the zeta potential is
(a) ±30 - ±60 (b) ±30 - ±40
(c) ±10 - ±30 (d) ±40 - ±60

764. Dynamic light scattering techniques preliminary is used for
(a) Size determination (b) Synthesize nanoparticle
(c) Shape determination (d) Characterization

765. Electron microscope is used to study the object with the size of
(a) >0.3 μm (b) <0.3 μm
(c) >2.0 μm (d) < 0.2 μm

766. Resolving capacity of electron microscope is
(a) 1.6 Å (b) 1.8 Å
(c) 1.4 Å (d) 1.2 Å

767. 2D image of the particles were produced by
(a) TEM (b) SEM
(c) AFM (d) EM

768. In SEM field emission gun required
(a) 10^{-10} torr (b) 10^{-6} torr
(c) 10^{-7} torr (d) 10^{-8} torr

769. Higher the atomic number of the atoms the more backscattered electrons are
(a) Emitted out (b) Bounced back
(c) Reflected in (d) Bounced

770. Covering the sample with a thin layer of conductive material with a device called
(a) Sputtering (b) Cutter
(c) Sputter coater (d) Vaporizer

771. The TEM samples are coated with ________ to improve a contrast in the microscope
(a) Pb (b) Au
(c) Ag (d) Al

772. Nanoparticles found wide applicability in fields such as
(a) Biomedicine (b) Diagnosis
(c) Wastewater treatment (d) All of the above

773. Nanoparticles are ________ times smaller than the width of a human hair
(a) 1000 (b) 10000
(c) 100000 (d) 1000000

774. ______________ are more uniform than the metal based nanoparticles and have unique material properties
(a) IVY nanoparticles (b) TiO_2 nanoparticles
(c) ZnO nanoparticles (d) Silica nanoparticles

775. According to the study, the nanoparticles substances damage the ______ functionality
(a) Bacteria (b) Fungi
(c) Virus (d) Parasite

776. The properties of nanoparticles are ___________
(a) Size dependent (b) Size independent
(c) Size neutral (d) None of the above

777. The semiconductor core shell nanoparticles are used for _________ purpose.
(a) Medical (b) Bio imaging
(c) Bio sensing (d) All of the above

778. The core shell classes of particles are also known as ___________ particles
(a) Smart (b) Small
(c) Great (d) All of the above

779. The class transition temperature is an important characteristic of polymer because below this temperature polymers are in state called __________ state.
(a) Solid (b) liquid
(c) Gas (d) Glassy

780. Nanoparticles may be defined as ______ colloidal particles
(a) Micron (b) Sub micron
(c) Mega (d) None of the above

781. Nanoparticles are the end products of wide variety of process
(a) Physical (b) Chemical
(c) Biological (d) All of the above

782. _______ are vesicular systems in which the drug is confined to a cavity surrounded by a unique polymeric membrane
(a) Nanotubes (b) Nano spheres
(c) Nano capsules (d) Fullerenes

783. ___________ are matrix systems in which the drug is dispersed throughout the particles
(a) Nanotubes (b) Nano spheres
(c) Nano capsules (d) Fullerenes

784. Nanoparticles can be prepared from a variety of materials, and ________
(a) Proteins (b) Polysaccharides
(c) Synthetic polymers (d) All of the above

785. Nanotechnology is the act of purposefully manipulating matter at the atomic scale, otherwise known as the _______.
(a) Nanometer (b) Nano scale
(c) Synthetic polymers (d) All of the above

786. Nanoparticles are dimensions between _________
(a) 1 to 10 nm (b) 10 nm and 1000 nm
(c) 1 nm and 1000 nm (d) 100 nm and 1000 nm

787. _________ is a wet chemical and promising method for this preparation of nano dimensional materials.
(a) Sol gel processing (b) Plasma method
(c) Pulsed laser method (d) Sono chemical reduction

788. An example of bottom up method is ________
(a) Physical synthesis (b) Chemical synthesis
(c) Biological synthesis (d) None of the above

789. Mechanical alloying is a solid state and powder processing technique involving repeated ___________ of blended powder particles.
(a) Cold welding (b) Fracturing
(c) Re-welding (d) All of the above

790. High energy ball milling is a mechanical deformation process that is frequently used for producing _________ metals in powdered form
(a) Crystalline (b) Nano crystalline
(c) Non crystalline (d) None of the above

791. High energy ball milling is effective in getting __________.
(a) Sparsely dispersed slurry (b) Well dispersed slurry
(c) Densely (d) None of the above

792. High energy ball milling, also called __________ can be used to reduce the grain size of materials from many micrometers to 2-20nm.
(a) Mechanical attrition (b) Physical attrition
(c) Chemical attrition (d) None of the above

793. Dimensions and tolerances of nanomaterial are in the range of ____________.
(a) 1 to 10 nm (b) 100 nm to 0.1nm
(c) 10 to 100 nm (d) 100 nm to 1 nm

794. ___________ of light is the optical properties of nanomaterial
(a) Absorption (b) Scattering
(c) Reflection (d) A and B

795. __________ is the electrical property of nanomaterials.
(a) Tunnelling current (b) Scattering
(c) Reflection (d) All of the above

796. Crystals contain internal interfacial defects known as ______________.
(a) Boundaries (b) Grain boundaries
(c) Sand boundaries (d) All of the above

797. __________ is force per unit area causing the deformation.
(a) Strain (b) Elasticity
(c) Stress (d) None of the above

798. _______ is a measure amount of deformation.
(a) Strain (b) Elasticity
(c) Stress (d) None of the above

799. The __________ is the constant of proportionality between stress and strain.
(a) Strain (b) Elastic modulus
(c) Stress (d) None of the above

800. ___________ is known as the father of biosensors.
(a) Leland C.Clark (b) Sumio Lijima
(c) Heinrich Rohrer (d) None of the above

801. A ___________ is an analytical device containing a biological recognition element.
(a) Carbon nanotubes (b) Fullerenes
(c) Indicator (d) Biosensor

802. The minimum amount of analytic that can be detected by a biosensor defines its __
(a) Limit of detection (b) Limit of assessment
(c) Limit of growth (d) Limit of blank

803. Electrochemical biosensor is based on the principle of change in ________
(a) Mechanical signal (b) Digital signal
(c) Electric signal (d) Energy signal

804. Optical biosensor is based on the principle of change in __________
(a) Absorbance (b) Reflection
(c) Refraction (d) Scattering

805. Piezoelectric biosensor is based on the principle of change in ___________
(a) Specific gravity (b) Density
(c) Volume (d) Mass

806. Thermal biosensor is based on the principle of change in _______________.
(a) Temperature (b) Pressure
(c) Humidity (d) Relative humidity

807. _________ is a temperature transducer
(a) Thermopile (b) Thermostat
(c) Thermistor (d) Thermocouple

808. In __________ transducer chemical energy is converted to light energy
(a) Optical (b) Current
(c) Pressure (d) Magnetic field

809. The basic electrochemical process includes __________
(a) Voltammetry (b) Potentiometry
(c) Conductimetry (d) All of the above

810. Non-catalytic biosensor is classified into two types namely ____________ sensor.
(a) Labelled affinity (b) Non labelled affinity
(c) Both a and b (d) None of the above

811. Calorimetric transducers are universally applicable in _________ sensor.
(a) Enzyme (b) Biological
(c) Biochemical (d) None of the above

812. Semiconductors are group of elements they has ______ electrons in their outer-most shell.
(a) One (b) Two
(c) Three (d) Four

813. The electrons in the outermost shell of an atom are called ________________.
(a) Valence electrons (b) Core electrons
(c) Valency electron (d) All the above

814. __________ material is chemically very pure and possesses poor conductivity.
(a) Intrinsic semiconductor (b) Extrinsic semiconductor
(c) N-type semiconductor (d) P-type semiconductor

815. Finite probability that an electron in the lattice will be knocked out from its position, leaving behind an electron deficiency called a ____________
(a) Hole (b) Hybridisation gap
(c) Doping (d) None of the above

816. Extrinsic semiconductor is an improved intrinsic semiconductor with a small amount of impurities added by a process known as __________.
(a) Hole (b) Hybridisation gap
(c) Doping (d) None of the above

817. Water-soluble derivatives of fullerenes inhibit the HIV-1 protein useful in fighting the _________ virus

(a) Ebola (b) HIV
(c) Zika (d) None of the above

818. Fullerenes can be added to polymers to create new fullerenes polymer blends with specific ________________ properties.
(a) Physical (b) Mechanical
(c) Electronic (d) All of the above

819. ________ are used for purification.
(a) Calixarenes (b) Toluene
(c) Acrylic (d) Polyester

820. Carbon nanotube is a tubular form of carbon with a diameter as small as ________.
(a) 1 nm (b) 0.1 nm
(c) 10 nm (d) 0.01 nm

821. Egyptians were using ____ to prepare make-up for eyes
(a) Nano aluminium (b) Nano copper
(c) Nano steel (d) Nano lead

822. The sword of Tipu Sultan was made of __________
(a) Nano lead (b) Nano aluminium
(c) Damascus steel (d) Pure iron

823. The thickness of a transistor is ____ nm.
(a) 50 (b) 90
(c) 2,000 (d) 5,000

824. Who had invented the famous 'Geodesic' dome structure?
(a) Eric Drexler (b) Buckminster Fullerenes
(c) Richard Smalley (d) Faraday

825. 10 nm _____ m
(a) 10^{-8} (b) 10^{-7}
(c) 10^{-9} (d) 10^{-10}

826. The size of a virus is ____ nm.
(a) 2 (b) 20
(c) 50 (d) 2000

827. The width of a typical DNA molecule is ____ nm.
(a) 1 (b) 2
(c) 5 (d) 10

828. 1 micrometre (micron) ______ m
(a) 10^{-9} (b) 10^{-8}
(c) 10^{-7} (d) 10^{-6}

829. 1 micrometre (micron) = ______ nm
(a) 1,000 (b) 100
(c) 10 (d) 0.01

830. What does 'F' stand for in AFM?
(a) Fine (b) Front
(c) Force (d) Flux

831. The surface area to volume ratio of a sphere with radius 1 cm is R_1 and that of a sphere with radius 5 cm is R_2. Then R_1 = ____ R_2.
(a) 3 (b) 1/3
(c) 5 (d) 1/5

832. The surface area to volume ratio of a cube with side 1 unit is R_1 and that of a cube with side 10 units is R_2. Then R_2 = ____ R_1.
(a) 1/10 (b) 10
(c) 1/100 (d) 100

833. The compressive strength of a nanotube _____ its tensile strength.
(a) is less than (b) is greater than
(c) is equal to (d) May be greater than

834. The hardness of a standard SWNT is ____ Pa.
(a) 63×10^6 (b) 25×10^6
(c) 25×10^9 (d) 25×10^{-9}

835. The bulk modulus of a standard SWNT is ________ that of diamond
(a) Less than (b) Greater than
(c) Equal to (d) Less than or equal to

836. How much current can be passed through 1 cm 2 cross-section of a metal nanotube?
(a) 10^{-9} A (b) 10^9 A
(c) 0.001 A (d) 1000 A

837. At room temperature, the thermal conductivity of a copper wire is ____ watt/(m.K).
(a) 3500 (b) 350
(c) 385 (d) 38.5

838. In radial direction, the thermal conductivity of a nanotube is ____ watt/(m.K).
(a) 3500 (b) 385
(c) 350 (d) 0

839. The thermal conductivity of an SWNT along length is ____ watt/(m.K).
(a) 35 (b) 350
(c) 385 (d) 3500

840. The size of a quantum dot is ____ nm
(a) 5 (b) 10
(c) 50 (d) 100

841. The capacity of a normal human eye to see the smallest object is ____ μm
(a) 10000 (b) 1000
(c) 100 (d) 10

842. Which of the following statement/s is are true?
(i) Volume to surface area ratio is very large for nanomaterials.
(ii) The cut-off limit of human eye is 10^{-5} m.

(iii) Hardness of a SWNT is about 63×10^9 Pa.

(iv) Carbon nanotubes are cylindrical fullerenes.

(a) All four (b) (ii) and (iv)

(c) (i), (ii) and (iv) (d) (ii), (iii) and (iv)

843. ____________ is a Short polymer between monomer and polymer.

(a) Oligomer (b) Bio polymer

(c) Natural Polymer (d) Linear

844. ____________ is a technique used for measuring the hydrodynamic size of a particle in suspension.

(a) Electron Backscattered Diffraction

(b) Dynamic Force Spectroscopy

(c) Dynamic Force Microscopy

(d) Dynamic Light Scattering

845. X-ray diffraction provide ______ information about the compound present in a solid sample

(a) Crystallinity (b) Chemical composition

(c) Stability (d) Porosity

846. Three dimensional x,y, z (Normal to the surface) images are obtained by

(a) Scanning Electron Microscope

(b) Scanning Tunnelling Microscope

(c) Transmission Electron Microscope

(d) Atomic Force Microscope

847. Quantum confinement is responsible for __________

(a) Discrete energy levels (b) Continuous energy levels

(c) Free electrons (d) Energy Bands

848. Typical infrared spectrum covers between __________

(a) 2.5 μm to 25 μm (b) 10 μm to 25 μm

(c) 6 μm to 30 μm (d) 7.5 μm to 20 μm

849. Force per unit area causing the deformation is ____________

(a) Hardness (b) Stress

(c) Strain (d) Yield

850. Optical biosensor is working based on ___________

(a) Light (b) Heat

(c) Electric Potential (d) Mass

ANSWER KEY

1. (d)	2. (c)	3. (d)	4. (a)	5. (d)	6. (d)	7. (c)	8. (b)
9. (b)	10. (c)	11. (a)	12. (a)	13. (b)	14. (a)	15. (c)	16. (c)
17. (b)	18. (a)	19. (c)	20. (a)	21. (d)	22. (b)	23. (a)	24. (d)
25. (c)	26. (c)	27. (d)	28. (d)	29. (a)	30. (c)	31. (d)	32. (d)

33. (c)	34. (a)	35. (b)	36. (c)	37. (d)	38. (a)	39. (c)	40. (a)
41. (a)	42. (d)	43. (a)	44. (b)	45. (d)	46. (c)	47. (a)	48. (a)
49. (a)	50. (a)	51. (d)	52. (a)	53. (b)	54. (a)	55. (a)	56. (a)
57. (a)	58. (d)	59. (a)	60. (d)	61. (a)	62. (b)	63. (a)	64. (a)
65. (d)	66. (a)	67. (d)	68. (a)	69. (a)	70. (d)	71. (a)	72. (c)
73. (a)	74. (b)	75. (b)	76. (c)	77. (b)	78. (a)	79. (a)	80. (b)
81. (b)	82. (c)	83. (a)	84. (c)	85. (d)	86. (a)	87. (c)	88. (c)
89. (c)	90. (b)	91. (d)	92. (d)	93. (b)	94. (a)	95. (d)	96. (c)
97. (b)	98. (a)	99. (b)	100. (a)	101. (a)	102. (a)	103. (a)	104. (c)
105. (d)	106. (a)	107. (d)	108. (a)	109. (d)	110. (a)	111. (b)	112. (c)
113. (d)	114. (d)	115. (d)	116. (b)	117. (c)	118. (a)	119. (c)	120. (b)
121. (c)	122. (c)	123. (c)	124. (b)	125. (a)	126. (d)	127. (b)	128. (c)
129. (d)	130. (d)	131. (b)	132. (d)	133. (c)	134. (d)	135. (b)	136. (b)
137. (a)	138. (a)	139. (b)	140. (c)	141. (a)	142. (d)	143. (a)	144. (c)
145. (b)	146. (d)	147. (c)	148. (a)	149. (b)	150. (a)	151. (b)	152. (c)
153. (a)	154. (a)	155. (a)	156. (b)	157. (a)	158. (b)	159. (d)	160. (a)
161. (c)	162. (d)	163. (c)	164. (d)	165. (a)	166. (b)	167. (b)	168. (c)
169. (a)	170. (d)	171. (a)	172. (d)	173. (a)	174. (d)	175. (b)	176. (a)
177. (c)	178. (b)	179. (c)	180. (b)	181. (d)	182. (c)	183. (d)	184. (c)
185. (a)	186. (b)	187. (a)	188. (a)	189. (a)	190. (a)	191. (a)	192. (a)
193. (b)	194. (b)	195. (b)	196. (a)	197. (b)	198. (a)	199. (b)	200. (b)
201. (a)	202. (d)	203. (b)	204. (b)	205. (a)	206. (c)	207. (d)	208. (b)
209. (b)	210. (b)	211. (b)	212. (a)	213. (c)	214. (a)	215. (a)	216. (b)
217. (b)	218. (b)	219. (b)	220. (b)	221. (a)	222. (b)	223. (b)	224. (b)
225. (a)	226. (d)	227. (c)	228. (c)	229. (d)	230. (b)	231. (a)	232. (b)
233. (b)	234. (b)	235. (d)	236. (a)	237. (b)	238. (d)	239. (d)	240. (d)
241. (c)	242. (a)	243. (b)	244. (a)	245. (a)	246. (c)	247. (d)	248. (b)
249. (a)	250. (a)	251. (b)	252. (b)	253. (b)	254. (a)	255. (a)	256. (d)
257. (a)	258. (d)	259. (a)	260. (b)	261. (b)	262. (b)	263. (d)	264. (d)
265. (b)	266. (b)	267. (b)	268. (d)	269. (d)	270. (d)	271. (c)	272. (d)
273. (a)	274. (d)	275. (c)	276. (a)	277. (d)	278. (a)	279. (c)	280. (c)
281. (b)	282. (b)	283. (a)	284. (a)	285. (a)	286. (b)	287. (b)	288. (c)
289. (d)	290. (d)	291. (b)	292. (a)	293. (b)	294. (c)	295. (b)	296. (a)
297. (a)	298. (c)	299. (d)	300. (c)	301. (d)	302. (c)	303. (a)	304. (b)
305. (a)	306. (a)	307. (d)	308. (a)	309. (b)	310. (a)	311. (d)	312. (d)
313. (a)	314. (a)	315. (a)	316. (c)	317. (a)	318. (a)	319. (d)	320. (a)
321. (a)	322. (a)	323. (a)	324. (a)	325. (a)	326. (a)	327. (a)	328. (a)
329. (a)	330. (a)	331. (a)	332. (a)	333. (a)	334. (a)	335. (a)	336. (a)
337. (a)	338. (b)	339. (b)	340. (c)	341. (d)	342. (c)	343. (c)	344. (d)

345. (b)	346. (b)	347. (a)	348. (a)	349. (b)	350. (d)	351. (a)	352. (a)
353. (c)	354. (d)	355. (c)	356. (a)	357. (a)	358. (a)	359. (b)	360. (a)
361. (a)	362. (d)	363. (a)	364. (a)	365. (c)	366. (b)	367. (b)	368. (a)
369. (c)	370. (b)	371. (a)	372. (c)	373. (b)	374. (c)	375. (a)	376. (c)
377. (b)	378. (c)	379. (b)	380. (b)	381. (a)	382. (a)	383. (b)	384. (a)
385. (a)	386. (a)	387. (b)	388. (c)	389. (d)	390. (c)	391. (b)	392. (a)
393. (a)	394. (b)	395. (c)	396. (d)	397. (d)	398. (d)	399. (d)	400. (d)
401. (d)	402. (d)	403. (d)	404. (d)	405. (d)	406. (d)	407. (d)	408. (a)
409. (d)	410. (d)	411. (a)	412. (b)	413. (a)	414. (a)	415. (c)	416. (a)
417. (d)	418. (b)	419. (c)	420. (b)	421. (a)	422. (b)	423. (b)	424. (a)
425. (b)	426. (c)	427. (d)	428. (a)	429. (c)	430. (a)	431. (d)	432. (a)
433. (d)	434. (c)	435. (d)	436. (b)	437. (d)	438. (a)	439. (c)	440. (a)
441. (d)	442. (a)	443. (b)	444. (d)	445. (c)	446. (c)	447. (c)	448. (c)
449. (b)	450. (b)	451. (b)	452. (c)	453. (b)	454. (b)	455. (b)	456. (c)
457. (a)	458. (d)	459. (b)	460. (a)	461. (b)	462. (b)	463. (b)	464. (b)
465. (a)	466. (b)	467. (a)	468. (c)	469. (a)	470. (b)	471. (d)	472. (d)
473. (b)	474. (c)	475. (b)	476. (a)	477. (a)	478. (a)	479. (b)	480. (b)
481. (a)	482. (d)	483. (a)	484. (a)	485. (b)	486. (b)	487. (a)	488. (a)
489. (a)	490. (b)	491. (a)	492. (b)	493. (b)	494. (a)	495. (a)	496. (c)
497. (d)	498. (b)	499. (a)	500. (d)	501. (a)	502. (a)	503. (a)	504. (b)
505. (a)	506. (a)	507. (b)	508. (a)	509. (b)	510. (c)	511. (b)	512. (c)
513. (a)	514. (a)	515. (b)	516. (b)	517. (a)	518. (b)	519. (a)	520. (d)
521. (a)	522. (d)	523. (a)	524. (d)	525. (a)	526. (b)	527. (d)	528. (a)
529. (d)	530. (d)	531. (b)	532. (a)	533. (a)	534. (c)	535. (d)	536. (b)
537. (c)	538. (b)	539. (a)	540. (b)	541. (c)	542. (c)	543. (b)	544. (c)
545. (d)	546. (b)	547. (c)	548. (c)	549. (b)	550. (a)	551. (b)	552. (c)
553. (d)	554. (b)	555. (a)	556. (a)	557. (b)	558. (b)	559. (b)	560. (a)
561. (c)	562. (b)	563. (c)	564. (a)	565. (c)	566. (a)	567. (d)	568. (b)
569. (c)	570. (d)	571. (d)	572. (a)	573. (d)	574. (c)	575. (d)	576. (b)
577. (c)	578. (b)	579. (d)	580. (a)	581. (a)	582. (b)	583. (b)	584. (b)
585. (b)	586. (b)	587. (b)	588. (a)	589. (c)	590. (a)	591. (c)	592. (c)
593. (c)	594. (c)	595. (b)	596. (b)	597. (c)	598. (c)	599. (a)	600. (b)
601. (c)	602. (b)	603. (a)	604. (c)	605. (a)	606. (b)	607. (a)	608. (c)
609. (c)	610. (a)	611. (a)	612. (b)	613. (a)	614. (b)	615. (d)	616. (d)
617. (b)	618. (a)	619. (b)	620. (a)	621. (d)	622. (b)	623. (d)	624. (d)
625. (d)	626. (b)	627. (b)	628. (a)	629. (a)	630. (a)	631. (d)	632. (b)
633. (c)	634. (a)	635. (a)	636. (c)	637. (b)	638. (a)	639. (d)	640. (b)

641. (a)	642. (b)	643. (a)	644. (a)	645. (c)	646. (b)	647. (c)	648. (a)
649. (b)	650. (c)	651. (a)	652. (a)	653. (b)	654. (d)	655. (c)	656. (c)
657. (c)	658. (b)	659. (a)	660. (d)	661. (a)	662. (b)	663. (c)	664. (b)
665. (a)	666. (b)	667. (c)	668. (d)	669. (b)	670. (a)	671. (b)	672. (b)
673. (a)	674. (c)	675. (a)	676. (c)	677. (d)	678. (a)	679. (c)	680. (b)
681. (d)	682. (b)	683. (d)	684. (b)	685. (c)	686. (b)	687. (b)	688. (c)
689. (c)	690. (c)	691. (a)	692. (d)	693. (a)	694. (a)	695. (a)	696. (a)
697. (b)	698. (a)	699. (d)	700. (a)	701. (a)	702. (d)	703. (b)	704. (a)
705. (b)	706. (d)	707. (c)	708. (a)	709. (a)	710. (c)	711. (a)	712. (b)
713. (a)	714. (d)	715. (a)	716. (a)	717. (a)	718. (a)	719. (b)	720. (a)
721. (a)	722. (b)	723. (a)	724. (b)	725. (b)	726. (c)	727. (c)	728. (d)
729. (c)	730. (d)	731. (c)	732. (b)	733. (a)	734. (b)	735. (a)	736. (a)
737. (a)	738. (b)	739. (c)	740. (b)	741. (c)	742. (d)	743. (c)	744. (c)
745. (b)	746. (a)	747. (b)	748. (d)	749. (b)	750. (c)	751. (a)	752. (b)
753. (b)	754. (c)	755. (a)	756. (b)	757. (b)	758. (d)	759. (a)	760. (b)
761. (c)	762. (b)	763. (d)	764. (a)	765. (d)	766. (a)	767. (b)	768. (a)
769. (b)	770. (c)	771. (a)	772. (d)	773. (c)	774. (a)	775. (a)	776. (a)
777. (a)	778. (a)	779. (d)	780. (b)	781. (d)	782. (c)	783. (b)	784. (d)
785. (b)	786. (c)	787. (a)	788. (b)	789. (d)	790. (b)	791. (b)	792. (a)
793. (b)	794. (d)	795. (a)	796. (b)	797. (c)	798. (a)	799. (b)	800. (a)
801. (d)	802. (a)	803. (c)	804. (a)	805. (d)	806. (a)	807. (d)	808. (a)
809. (d)	810. (c)	811. (a)	812. (d)	813. (a)	814. (a)	815. (a)	816. (c)
817. (b)	818. (d)	819. (a)	820. (a)	821. (d)	822. (c)	823. (b)	824. (b)
825. (a)	826. (c)	827. (b)	828. (d)	829. (a)	830. (c)	831. (c)	832. (a)
833. (a)	834. (c)	835. (b)	836. (b)	837. (c)	838. (d)	839. (d)	840. (a)
841. (d)	842. (d)	843. (a)	844. (d)	845. (a)	846. (d)	847. (a)	848. (b)
849. (b)	850. (a)						

2

Glossary – Nanotechnology

A

Activated carbon: Adsorbent made from carbonaceous materials such as coal or charcoal, which is chemically or physically processed to enhance purification.

Activation energy: It is the initial energy required by the precursor materials to transform into products. Certain parameters like heat, cold and light can initiate chemical reactions. For example, condensation and nucleation process.

Adenosine triphosphate: is a molecule that transports energy within living cells. It is also called as energy currency of cells.

Adhesion: it is the ability of two different particles or surfaces to stick together.

Adsorb: refers to the adhesion of liquid or gas molecules to the surface of a solid particle.

Aerogel: is a porous, lightweight and synthetic solid material derived from a gel in which the liquid component of the gel is replaced with a gas.

Aerosol: is a colloidal suspension of very small solid or liquid particles in gas or atmosphere.

Affinity-contact printing: captures a ligand or a target molecule selectively from a medium **Agglomeration:** Small unstable molecules tend to colloid with one another and grow into larger units. When an unstable nanoparticles collide in a suspension, agglomeration takes occurs and releases them as a printed pattern on a substrate.

Aggregation: is a collection of single or varied things into a group or cluster.

Albumin: it is a globular protein made by the liver and found in blood plasma.

Alginate: is a natural anionic polymer derived from the cell wall constituents of brown seaweed. It is a popular gelling and thickening agent.

Alkali metals: they are periodic table Group 1 (Ia) elements, such as sodium and potassium that react with water to form strong bases or alkalies.

Alkoxysilanes: they are chemically reactive silane compound with alkoxy moieties. Commonly used as a crosslinking agent and for hydrophobic surface coatings.

Alumina: is also known as aluminium oxide (Al_2O_3). It is typically found in rocks and clays, which are also extensively used as ceramic materials.

Amino acids: they are composed of nitrogen, hydrogen and carboxyl groups and form the basic units of proteins.

Aminopropyltriethoxysilane: it is a salinizing reagent used to immobilize amine groups to silica or glass surfaces.

Amorphous oxides: it is a non-crystalline solid material in which the atoms or the molecules are not arranged in a definite pattern in a crystal lattice.

Amphiphilic molecule: is one molecule that has both the hydrophilic (polar) and hydrophobic (non-polar) moieties. They are used for self-assembly and surface modification applications. Surfactants, detergents and soaps are few examples of amphiphilic compounds.

Angstrom: The Symbol of Angstrom is (Å) and it represents 10^{-10} meter or 0.1 nanometer.

Anionic: atoms or molecules that are negatively charged or substrates with negatively charged surfaces.

Anisotropic: materials that exhibit different property at different orientations or direction-dependent properties. Example: Diamond, glass and crystals.

Anthropogenic pollution: pollution or any negative effects on the environment brought on by human activity.

Antibacterial: any material or drug that can kill or inhibit the growth of bacteria.

Anticancer: any material or drug that has the property to kill or slow down the growth of cancerous cells.

Arc discharge method: It is a method to prepare carbon nanotube. When an electric current is applied to two graphene electrodes, plasma is produced in which carbon is vaporised and deposited as carbon nanotubes.

Arsenic (As): It is a heavy metal pollutant. Long-time exposure to arsenic can cause arsenicosis and cancer.

Artificial cells: it is a synthetic cell or any engineered particle that mimics the functions of the biological cells.

Artificial intelligence: it is referred to systems or machines that can mimic the human intelligence to perform tasks.

Assembler: Assemblers are Nano devices and nanoparticles that can self-assemble into a functional design or replicate themselves. It is also known as Nano assembler. (or) It is a computer program that can interpret the software program language into machine language to run the instruction.

Atom: it is the smallest unit of the element that can exist alone or in combination.

Atomic force microscope: it is a high resolution scanning probe microscopy. It uses cantilever and probe tips to scan the surface topography of a material at atom level.

Atomic layer deposition: it is a vapour phase technique used to deposit atoms or molecules as thin films onto a substrate.

Au nanoparticle: Gold (Au) nanoparticles exhibit surface plasmon resonance and have a unique size dependent optical property. As a result, it is extremely useful in bio-imaging as well as therapeutic applications.

Auger electron Microscopy: It is an electron microscopy technique that detects the auger electrons (electrons from the inner shell) emitted from a sample to obtain high resolution surface image.

B

Band gap: is also called as an energy gap, a region in which no electronic structure is present. It is described as the difference in energy levels between an electron's valence band and conduction band.

Beam pen lithography: it is a photo-induced lithographic technique used to pattern nanoscale features on various surfaces including integrated circuits and diagnostic devices.

Bimetallic nanoparticles: it is the combination of two different metal or metal oxide nanoparticles that exhibit new and enhanced properties than their individual counterparts.

Bioaccumulation: is referred as the gradual accumulation of chemicals in living organisms over time, resulting in higher concentrations of those chemicals in the body because the accumulation rate is faster than the elimination rate.

Biochip: The concept of a lab on a chip, in which numerous biochemical tests can be performed in parallel over a chip to screen for disease-specific antigens or antibodies.

Biocompatibility: it is defined as the ability of any foreign material to be compatible with living cells without triggering any immune response.

Bio composites: it is a composite of two or more materials derived from natural sources.

Bio distribution: it is the tracking or detection of the movement of chemicals inside living organism from one place to another.

Biofilms: It is a thin film formed by bacterial cells embedded in extracellular matrix and adhered to a solid surface.

Bio fouling: It is the process by which microbes, algae, and plants colonise and accumulate on a submerged or wetted surface, forming a biofilm. Bio fouling occurs on submerged pipes, ship surface, and underwater devices.

Biogenic nanoparticle: the nanoparticles that are synthesized from a biological sources such as plants, bacteria, fungi and actinomycete

Biological oxygen demand (BOD): It is a water quality testing parameter that indicates an aquatic organism's oxygen demand or requirement level in order to survive. The BOD level rises in a polluted aquatic ecosystem.

Bio magnification: is the transfer and accumulation of toxic chemicals at higher concentrations from organisms at lower trophic levels to those at higher trophic levels via the food chain.

Biomarker: it is a biological marker or an indicator that can predict the normal and the abnormal health state of an individual such as blood pressure and cholesterol levels.

Biomedical Nanotechnology: It is a rapidly developing field in which nano technological applications are used in current therapeutic methods in the biomedical and health care fields to diagnose and treat a wide range of diseases.

Bio-MEMS: is an abbreviation for Biomedical Micro electro mechanical systems. it is a device constructed by incorporating a biological or chemical component to detect and treat the diseases

Biomimetic: it refers to as designs or systems developed by mimicking the nature.

Biomolecule: they are biological molecules produced by living organisms such as polysaccharides, proteins, amino acids and nucleic acids

BioNEMS: It is abbreviated as biomedical nano-electromechanical systems. It is a nanoscale device with ultra-small dimensions. For example, Nano-electrochemical sensor is a BioNEMS, which has nanoparticle coated electrodes and nano recognition materials to detect the biomolecules like protein, DNA and whole organism.

Biopolymer: natural polymer derived from biological origin such as plants and microbes

Bioreactor: A fermentor is a large cylindrical vessel or tank with a suitable environment that facilitates the production of large volumes of cell culture, such as plant, microorganism, and human cells.

Bioremediation: It is an eco-friendly pollution remediation approach that uses micro-organisms to degrade the pollutant and transform into less harmful substances. Bioremediation can also used along with other

Biosensor: refers to a device that contains a biological component, such as an enzyme, which interacts with an analytic to produce biological or chemical responses, which are then converted into electrical signals and measured.

Bio surfactants: They are the amphiphilic molecules that are naturally derived from micro-organisms such as bacteria, yeast and fungus.

Biosynthesis: also referred to as "green synthesis," is a process for synthesizing nanoparticles from living organism like plants and microbes.

Biotin-avidin system (BAS): Avidin is a protein that binds strongly to biotin molecules; this is a very specific protein interaction. The affinity between biotin and avidin is used in a variety of applications such as affinity chromatography, sensors, and bio-imaging.

Block co-polymers: Co-polymerisation is the chemical crosslinking or covalent attachment of two or more polymers. Block co-polymers are polymers that have two or more distinct homopolymer units called blocks that are chemically combined.

Blood brain barrier: is a network of closely spaced semipermeable endothelial cells that helps to keep harmful substances away from reaching the brain.

Blue goo: This is a hypothetical scenario in which nanorobots are used for beneficial technology, in contrast to grey goo where nanorobots replicates to be a destructive.

Braggs Law: X-ray diffraction spectroscopy (XRD) determines the crystal structure of a material using Braggs Law and the equation $n\lambda=2d \sin\theta$, where (λ) is the wavelength of the X-ray beam that hits on a crystal surface and (2θ) is the angle of incidence and the angle of diffraction, (d) is the interplanar distance in the crystal lattice and (n) is the integer.

Bottom-up approach: it is a method to synthesis larger nanomaterials by combining smaller atoms and molecules.

Brownian motion: also called as a Brownian movement, it is a random zigzag motion of a particle that is suspended in a medium.

Buckyball: also known as buckminsterfullerene is a type of fullerene (C60), an amorphous carbon connected in a pattern of hexagons and pentagons to form spherical closed cage shape resembling a soccer ball.

C

Cadmium (Cd): it is a heavy metal and highly hazardous contaminant that affects human and other living organisms when exposed it over time. Cd is produced from smoke, batteries and industrial waste.

Computational Nanotechnology: It is a field of nanotechnology that uses computer based modelling to study the interactions and working of molecules and devices at nano scale level.

Cantilever: it can be a horizontal projection of a solid material with one end fixed and the other is flexible. Cantilevers in AFM are made of silicon or silicon nitride material and are attached with a tip. They are used to scan the surface of the specimen.

Conduction band: It is the highest energy band in the electronic orbit, where the electron from the valence band gets energized and jumps to reach the conduction band to conduct electricity.

Carbon: it is a one of the abundant element found in Earth's crust. In Latin the word carbon means "Coal". In the period table, it is classified as the non-metallic chemical element with the symbol C and atomic number 6.

Carbon nanotubes (CNT): also referred as "backytubes", are one dimensional nanomaterials in cylindrical or tube forms. They are made from carbon allotropes such as graphite sheets.

Catalyst: it is defined as the substance that can alter the rate of the reaction without itself being consumed in the process.

Cell: The smallest, basic unit of life that is responsible for all the life's process.

Cell adhesion: it is the binding of one cell to the other cell to facilitate cell-cell interaction using adhesive molecules

Cell membrane: it is the outermost part of the cell, made up of semipermeable lipid bilayers that protects the cytoplasm and other intracellular components while regulating the passage of other substances into the cell.

Cell recognition: is the ability of a cell to produce specific responses following the cell's interaction with a neighbouring cell or with a foreign material.

Cellulose: It is a complex and commonly found polysaccharide in nature with a chemical formula $C_6H_{10}O_5$ that serves as the main organic building block of the cell walls of plants and other micro-organisms like algae. Cellulose also supports the growth of bacterial biofilm.

Chemical vapour deposition (CVD): is a bottom-up method for making nanoparticles. This process involves heating the volatile precursor material to break it down into the desired elements, which are then deposited as thin films over the substrate under vacuum condition.

Chemiluminescence: it is referred to as the emission of electromagnetic radiation or light from a substance after a chemical reaction. When an electron is excited from its ground state to its excited state during a chemical reaction, the excess energy of the electrons is released as light (luminescence) when it returns to its ground state.

Chemisorption: it is the process in which solid, liquid or gaseous substances (adsorbate) are adsorbed to a surface of another substance (adsorbent) through chemical reactions. Chemisorption typically results in the formation of strong covalent bond between two substances and making it a very specific and an irreversible process. The rate of the chemisorption reaction can be accelerated by increasing the temperature and by using a metal catalyst.

Chlorofluorocarbon (CFC): It is a synthetic gas containing chlorine, fluorine, and carbon substances that is released by air conditioners, refrigerators, and fire extinguishers etc. It has the potential to deplete the ozone layer.

Chromium (Cr): It is a heavy metal that is most commonly found in the +3 and +6 oxidation states. Cr+6 is highly toxic and is classified as an irritant, health hazard, and environmental hazard.

Clay minerals: are Alunimosilicate minerals containing elements such as aluminium, silicon, iron, magnesium, and other alkali and alkaline earth metals. kaolinite, montmorillonite, and halloysite are few examples of clay minerals.

Colloids: It is a two-phase system in which ultrafine insoluble particles in the range of 1-1000 nm or gas particles (dispersed phase) are evenly suspended throughout the another medium, such as a fluids (continuous phase). There are four types of colloids: emulsion, gel, aerosol, and foam.

Composite: It is a material synthesized from two or more materials with different physical or chemical properties. The property of the composite material is unique and different from their parent material. Various metals, metal oxide and ceramic materials are used to form composites and are used in application like reinforcement of concrete, plastics, fibres and etc.,

Computational chemistry: It is an advanced field of chemistry that analyses and predicts all possible outcomes form an experiment using computers programmes with theoretical chemistry inputs. Computational chemistry is sometimes used to design structures and binding sites in molecules to improve their interaction with other molecules.

Conductivity: is the amount of heat or electric current that can pass through a material. There are many metal conductors like copper, silver, and gold, whose resistance are low and allow free electron flow.

Confocal microscopy: is a high-resolution imaging technique that studies the morphology of a material by using laser light as a source of illumination. The point illumination principle is used in confocal microscopy, where the material to be evaluated is placed at the plane of focus, darkening the background, and the laser beam is specifically focused on the fluorescently labelled material of interest alone, yielding a contrast micrograph of the material. Biological samples such as fixed or live tissues and other microorganisms are typically analysed.

Copolymerization: is process in which the combination two or more different types of monomers are polymerized simultaneously. This process improves the physical and mechanical properties of monomers in the combined copolymer form such as tensile strength, thickness, solubility, crystallinity, and wettability.

Co-precipitation: It is a chemical method of synthesizing nanoparticles in which the precursor materials solubilise the solvent at a specific temperature and pressure to precipitate nanoparticles.

CRISPER: It is a biotechnological tool used to modify or edit gene sequence. It is also called as base editors. It can be use to modify defective genes that causes disease like autoimmune disease, neurological disorders, cardiovascular disease and cancer.

Crystal: is a solid substance whose atoms or molecules are uniformly arranged in a specific pattern in the crystal lattice. Crystals are divided into various types based on their unique internal atomic structures, like cubic, hexagonal, monoclinic, triclinic, orthorhombic, tetragonal, and trigonal structures.

Crystallography: is an analytical technique to study the crystal structure, like the atom position, three-dimensional atomic structure, atomic bonding, bond length and angles. The information about the crystal structure can be used to determine the properties of the crystal like why the Diamond is very hard and why salt is soluble in water. X-ray diffraction (XRD) crystallography is one such analytical technique used commonly to study the crystal structure.

Curcumin: It is a yellow chemical substance present in plants like turmeric and ginger. It has antioxidant properties and are used in food, supplement and cosmetics.

Cytotoxicity: refers to any substance that is toxic to cells and results in cell damage or cell death. Cytotoxicity studies determine the toxicity of a chemical substance to cells. These studies are critical in many fields, including nanotechnology, to determine the potential toxicity of newly developed nanomaterials. Cytotoxicity studies can be performed on a variety of cell lines and animals through in vitro and in vivo toxicity assays.

D

Damascus steel: It is well-known steel from the 17th century that was used in weapon blades. is made of wootz steel, which is manufactured in India. The steel sword had a high carbon content of around 1.8% and was made of an iron-carbon alloy, but steel swords with this carbon content are typically brittle. Surprisingly, Damascus steel was both hard and malleable. When the blades were examined, it was discovered to have cementite nanowires and carbon nanotube structures, which could have contributed to its unusual hardness.

Data mining: is the process of discovering patterns and information in large amounts of data. It essentially entails exploring the data, modelling it, and evaluating it in order to solve problems based on the value of the work.

De Broglie wave length: The de Broglie concept proposes that all matter can be both particle and wave in nature. It is a crucial concept in quantum mechanics. It is an object's wavelength in relation to its momentum and mass.

Debris: is a term that refers to remains, trash, dirt, or waste that is dispersed.

Deflection: is the term used to describe the change in an object's velocity after it deviates from its original position due to contact or hitting with a plane surface. This deflection principle is commonly used in analytical techniques such as AFM, where the beam is deflected over a cantilever tip to scan the surface topography of the nanomaterial.

Delamination: is the separation of a material into layers that occurs when there is weak interaction between layers. Materials with multiple layers, such as graphite oxide sheets, must be delaminated in order to produce a single layer of nanosheet.

Dendrimer: is a nano-sized three-dimensional polymer, with highly branched symmetrical structure and an atom in the centre, similar to a core-shell like structure. Since dendrimers also have amphiphilic moieties in their structure, they can be used as nano-carriers to load both hydrophobic and hydrophilic drugs in large quantities.

Detector: it is a device which detects the presence any substance like target material, metals, electromagnetic waves, pollutants, smoke, fire and so on.

Detectors are used as device in nanosensors and also in analytical instruments like X-ray diffraction spectroscopy and Scanning electron microscopy

Dialysis: is the separation of particles or colloids suspended in a solution based on the diffusion difference rate through a semipermeable membrane.

Dialysis medically, is a treatment used to filter out excess fluid in the body and purify the blood when a person's kidneys fail.

Dichlorodiphenyltrichloroethane (DDT): is an organochlorine insecticide used to control agricultural pests, household pests, and mosquitoes that transmits malaria. DDT is classified as a persistent organic pollutant (POP), and its use is restricted in many countries due to potential environmental and human health hazards.

Dielectric: a material with the property of an insulator; it lacks free electrons and thus is a poor conductor of electricity; however, when placed in an electric field, it becomes polarised and transmits electric force rather than conducting it. This phenomenon occurs when the positive charge of a material is directed to one side and the negative charge is directed in the opposite direction under the influence of an electric field.

Diffraction scanning colorimetry (DSC): it is a thermo-analytical instrument used to measure the heat transitions of a material when exposed to different temperature over a period of time. DSC measures the phase transitions like melting, thermal stability, crystallization, polymorphism, purities, compositional changes, decompositions and so on.

Dimethylformamide (DMF): is a colourless organic solvent with a molecular formula $(CH_3)_2NC(O)H$ that is widely used to disperse and stabilize metal nanoparticles.

DMF also act as a strong reducing agent, reducing $Ag+$ ions to form silver nanoparticles. As a result, it can be used as both a reducing agent and a solvent.

Dip-pen nanolithography: also known as scanning probe lithography is a bottom-up nanofabrication technique that uses the probe tip of an atomic force microscope to create high-resolution material and molecular patterns on a substrate, also known as "molecular inks."

Directed Self-assembly: is a bottom-up method of nanofabrication, where molecules or materials automatically self-assemble into accurate designs or patterns of interest under the influence of external forces such as magnetic field, electrostatic force, gravitational force, etc.

Dispersability: is the ability of a particle to be evenly distributed throughout a solution. For instance, nanoparticles like carbon nanotubes, iron, and silver can disperse to form a homogenous suspension in water or organic solutions. The size and concentration of nanoparticles are typically determined after dispersion, making it a crucial nanoparticles property.

Disruptive technology: is a new technology that replaces old or more established technology. It opens up new opportunities for business operations by creating innovation in products, models, and networks. Examples of disruptive technology include 3D printing and artificial intelligence (AI).

DNA: abbreviated as deoxyribonucleic acid, is a polynucleotide chain with a double helical structure that carries genetic information that code for growth, development, and body functions.

DNA origami: is the folding of long, single-stranded DNA to create fascinating 2D & 3D nanostructures such as smileys, pyramids or any template design of interest. It is used for the construction of devices, nanobots, enzyme-substrate interaction and drug delivery.

DNA recognition: is a process in which molecules, such as proteins are modified to identify and bind to the target DNA sequence that specifically codes for the protein, and self-assembles to form a protein-DNA complex for various therapeutic purposes.

Dopamine: is a "chemical messenger" or neurotransmitter produced in the brain that plays an important role in signal transmission between nerve cells and hormone regulation. The amount of dopamine levels in the brain affects the person's mood, behaviour and organ functions like heart rate, kidney function and blood vessel function.

Dopants: are the impurities added in trace amounts to a material to improve its physical and chemical properties. This is referred to as doping. For example, dopants such as arsenic, boron and gallium are added to semiconductor materials such as silicon to improve their conductivity.

Dose: is the concentration or quantity of a drug given to a person or animal during the treatment process, and it is determined by factors such as age, gender, and body weight.

Dose-response curve: is a graphical representation on the effects (response) of drugs given to a person or animal at a particular drug concentration (dose) over a period of time. The dose-response curve is sigmoidal curve indicating the dose at the X-axis and the response at the Y-axis. Initially, the drug at the lower concentration is insufficient to cause any response, but at the increasing dose, there is an increase in the response and it reaches the maximum efficacy at a point.

Doxorubicin: it is a chemotherapeutic drug used to kill or slow down the growth of the cancerous cells. It is used to treat many types of cancers including breast cancer, bladder cancer and etc.

Drug delivery system (DDS): is a technology that uses carrier materials to deliver drugs at the target site, to increase the efficacy of the drugs and to reduce the harmful side effects. Many nanomaterials like Magnetic, gold, carbon nanoparticles, polymer and lipid nanostructures are used as nano-carriers to deliver the drugs.

Dry etching: is a method of carving a pattern into a semiconductor material such as silicon dioxide thin films. This technique, also known as plasma etching, involves cutting out a masked surface on a semiconductor material to create a pattern by bombarding it with reactive ion plasma such as fluorocarbons.

DWNT: stands for double walled nanotube, is a one-dimensional nanotube structure. Arc discharge methods can be used to create single, double, and multi-walled carbon nanotubes.

Dyes: it is a chemical compound used to color materials like paper, fabrics, cosmetics, foods and so on. Dyes are produced naturally from plant sources as well as synthetically from chemical compounds like benzene derivatives.

Dynamic Light Scattering (DLS): is a technique used for measuring the hydrodynamic size of a particle in suspension. DLS is based on the Brownian movement of suspended particles, where the smaller particles diffuse faster and scatter more intense monochromatic light than larger particles, which move slowly and scatter less intense monochromatic light. Therefore, the changes in the monochromatic light intensity are correlated with the size of the particle.

E

Ecology: The study of how an organism interacts with its surroundings, both living (biotic) and non-living (abiotic) factors, to determine how vital the environment is to the organism's survival and growth.

Ecosystem: An area or a unit space where the living organisms like plants and animals linked with the each other and with the non-living surroundings like soil, water and air through nutrient cycle, energy flow and etc.

Eco tone: a transition zone or boundary between two ecosystems that are rich in biodiversity, also known as a biodiversity hotspot. Mangrove forest, for example, serves as an ecotone between the marine and terrestrial ecosystems.

Ecotoxicology: a field of study that integrates toxicology and ecology to foresee and analyse the toxic effects that chemicals and other contaminants have on both organism and environment.

Ecotype: also called as ecospecies, is a population of species that have evolved to exist solely in a specific habitat or to a particular environmental condition.

Energy Dispersive X-ray: is a method to identify the elemental composition of a material in conjugation with electron microscopy. EDX utilises the topographic data produced by electron microscopes like SEM and TEM to analyse the materials composition at a particular point (region).

Elasticity: The ability of solid materials to regain their previous shape and size after being deformed. It measurement made to find a material's strength or the breaking point (strain) to withstand a load (stress).

Electrochemical reaction: It is a spontaneous redox reaction that occurs in an electrochemical cell when electrons flow from one electrode to another via electrolyte to produce electricity.

Electrode: it is a conductor or a semiconductor material present in the electrochemical cell or batteries, it is a point where the electrons transfer takes place and therefore the electricity generates. There are two types of electrode cathode and anode.

Electroluminescence (EL): It is a property of semiconductor material to emit light under an influence of an electric field. During this process, electrons are excited to higher energy states and the recombination of electron-hole pairs at the ground state results in the release of excess energy in the form of light or photons.

Electromagnetic (EM) waves: are oscillating waves formed when an electric and magnetic field overlap. EM waves carry energy at varying intensity and move through space. For example, Radio waves, microwaves, infrared light, visible light, ultraviolet light, X-rays, and gamma rays are the seven forms of electromagnetic waves.

Electron Beam Lithography: it is a bottom up approach to fabricate fine structures of nanomaterial. In this method, the electron beam is focused onto a surface of a specific pattern coated in electron sensitive film, and through the etching process the desired nanostructures is generated.

Electron diffraction: This process is similar to X-ray diffraction, but instead of X-rays, an electron beam is used to identify the crystal structure of the material through the diffracted electron pattern. This technique is used in transmission electron microscopy (TEM) to analyse the material's crystal structure.

Electron microscopy: is a type of advanced microscopy that uses high-energy electron beam to investigate the morphology of very small objects and generate high-resolution two- and three-dimensional images. SEM (scanning electron microscopy) and TEM (transmission electron microscopy) are two types of electron microscopy.

Electron tunnelling: is also known as quantum tunnelling, electrons pass through a potential barrier as a wavefunction. This is impossible according to classical mechanics if one considers the electron to be a particle. This electron tunnelling phenomenon is used in scanning tunnelling microscopy (STM) to obtain high resolution images.

Electronic defects: holes or oxygen vacancies that arise due to the unavailability of electrons in the atomic structure. In an electric field, electrons move to the conduction band form the valence band, generating holes, and the holes in the valence band move to the opposite direction of an electron and thus electron defect bond is generated.

Electronic Nose: often known as **E-Nose**, is an electronic gas sensor developed with an array of metal oxide materials to sense specific or complex odours or gases. For example, the detection of specific diseases based on the presence of specific volatile organic compounds in a person's breathe.

Electrons: represented as e^- or β^-, are subatomic particles with a negative charge. Electrons always revolve around an atom's nucleus, and the electron in the valence shell binds to the neighbouring atoms to create a molecule.

Electrophoresis: this technique separates biological macromolecules using electric current. Macromolecules such as DNA, RNA, and proteins are separated on a gel based on their surface charge, size, and affinity. For example, in Agarose gel electrophoresis, the negatively charged DNA move towards a positive electric terminal to get separated, and larger DNA fragments move slowly and smaller DNA fragments move faster across the pores of the gel.

Electrospinning: is a nanofiber synthesis technology in which a polymer solution from a syringe is charged and dragged out of a needle tip by an electric force, forming a Taylor cone shaped projection, and the solvent solution evaporates, leaving the dry polymer to deposit on a collector to form thin nanofibres.

ELISA: stands for enzyme-linked immunosorbent assay, in which proteins, peptides, and hormones are immobilised on well plates and the antibodies coupled with enzymes are allowed to interact. After the specific antigen-antibody reaction, changes in the activity of reporter enzymes are measured using an ELISA plate reader at a specific absorbance.

Emulsion: It is a uniform blend of two or more liquids which usually don't mix and phase separate under normal condition. For example, Oil in water emulsion (o/w) and water in oil emulsion (w/o) are two types of emulsion systems.

Encapsulation: is the technique of encasing a bioactive material (core) inside a carrier material (shell) in order to preserve the physicochemical properties of the bioactive material.

Engineered Nanoparticles (ENPs): are nanomaterials (less than 100 nm) with size, shape, and surface chemistry modified to fit in certain applications like drug delivery, nanosensing, photocatalysis, etc.

Entropy: it is thermodynamics theory to assess the degree of disorder or uncertainty in a system, which denotes the system's inability to convert thermal energy to mechanical energy or to do work.

Environmental ethics: is concerned with how human actions affect the environment and develop morals and policies to value both living and non-living components of the environment.

Environmental risk assessment (ERA): is a method that assesses the harmful effects or risk associated with specific human activities on the environment, such industrial activity,GHG emissions, and hazardous material release.

Enzyme: are proteins that work as a biological catalyst in the body to regulate certain chemical reactions. The enzymes interact with a substrate molecule like a lock and key model and convert it into different products.

Eric dexler: He is a nanotechnologist and engineer who specialises in molecular nanotechnology. In 1986, he published the book Engines of Creation- The Coming Era of Nanotechnology, which describes the Feyman speech, nanosystems and molecular machinery.

He is a nanotechnologist and engineer who specialises in molecular nanotechnology. In 1986, he published the book Engines of Creation- The Coming Era of Nanotechnology, which describes the Feyman speech, nanosystems and molecular machinery.

Eutrophication: is a phenomenon that occurs when there is an algal bloom in a water body, followed by a decrease in dissolved oxygen as a result of a steady increase in minerals and nutrients, such as nitrogen and phosphorus from nutrient-rich settings.

Ex situ remediation: It is a technique in which the contaminated soil or water is removed from the site of contamination and treated at the point of use or away from the contaminated area.

Extra cellular Matrix (ECM): Other than cell components like fibrous proteins such as collagen, enzymes, and glycoproteins that add structure and support to the cells. ECM is a key component in the adhesive mechanism of bacterial biofilm formation, where it attaches to the cell and the surface.

F

Fabrication: is the process of manufacturing or designing materials from raw resources. Nanomaterials are synthesised using bottom-up and top-down techniques.

Faraday (F): also referred to as the Faraday constant, is the measure of the electric charge that each mole of an electron carries. This unit is named after a great scientist Michael Faraday, who worked in electromagnetism.

Fauna: wildlife of a particular region, which includes animals, birds, insects, reptiles, fish and other marine life.

Food and Drug Administration: is a federal agency in the United States whose mission is to ensure the safety and efficacy of goods such as food, medicines, cosmetics, and devices that emits electromagnetic radiation.

Femtotechnology: it is the study of atoms or matter at femto scales (10^{-15} m), which are smaller than nano (10^{-9} m) and picometer scales (10^{-12} m). This technology is currently considered hypothetical, since advanced instruments must be developed to explore the atomic energy levels at femto scales.

Fermi level: is the highest energy level an electron can occupy in a semiconductor material at absolute zero temperature. The Fermi level of electron resides between the bandgap or forbidden energy gap, which is the gap between valence band and conduction band of the semiconductor material.

Ferrites: are soft magnetic materials with ferrimagnetic characteristics that have magnetic domains aligned in the opposite direction to the magnetic field and thus can be easily magnetised and demagnetized. Metal oxide like magnetite (Fe_3O_4) is an example of ferrite.

Ferroelectrics: is the ability of a non-conducting material or dielectrics to exhibit spontaneous electric polarization when placed in an electric field, also known as "switching" mechanism, and the polarization remains within the material even after the electric field is removed.

Ferrofluid: it is a colloidal dispersion of nano-scale ferromagnetic material that is attracted to magnetic poles, also referred to as magnetic fluid.

Ferromagnetic materials: are those that are highly attracted to magnets and have a tendency to display magnetic characteristics when subjected to magnetic fields. Certain metals like Iron, Copper, Nickel and Gadolinium are examples of ferromagnetic materials.

Fertilizers: are any inorganic or organic compounds that are used to replenish the soil with nutrients in order to increase plant growth and productivity. Urea and ammonium sulphate are examples of nitrogenous fertilizers.

FESEM: stands for Field Emission Scanning Electron Microscopy is a high resolution electron microscopy technique that uses a potential gradient to emit electrons from a field emitter gun (FEG) to scan the surface of the target material, obtaining the topographic image and elemental information.

Feynman: Richard Philips Feynman is a theoretical physicist who won the Nobel Prize in Physics in 1965 for his work in quantum electrodynamics. His talk at the American Physical Society meeting on "There are plenty of rooms at the bottom," marked a new beginning for the area of nanotechnology, and he is widely called as the father of nanotechnology.

Fiber optics: is a technology that uses light pulses to transport information across vast distances using flexible, thin, and transparent cables composed of plastic, glass, or silica. Fiber optics is used to replace copper wire in telecommunications and internet connections.

Field effect: refers to the changes in the electrical conductivity of a material by the influence of an external electric field.

Field emission: is a technique in electron microscopy that uses a strong electric field to generate electrons from a tungsten electron gun source. This approach is also called as "cold emission", as the tungsten is not heated to generate electrons; instead, a potential difference is created to generate electrons.

Field-effect transistor (FET): a three-terminal type transistor that uses an electric field to control the flow of current from the source to the drain terminals via a gate terminal in a semiconductor material.

Filler materials: any substance, often ceramics such as calcium carbonate, silica, and clay, is utilised as a filler material or matrix to reinforce the attributes of the base materials such as strength, density, and shape.

Filtration: is a physical process that uses a medium or semipermeable membrane to separate undesirable or harmful gas, solid or liquid substances from a mixture.

Flocculation: is an aggregation process that takes place when colloidal particles in suspension come together to form large clumps and ultimately settles down as flocs in the suspension.

Flora: is any organism that is capable of producing its own food, it includes the plant life such as trees, shrubs, flowers and fungi.

Flue gas: also known as exhaust gas, is the mixture of gases that escape into the environment as a result of the combustion process through a pipe or the flue. Flue gases include gases such as sulphur dioxide, carbon dioxide, and carbon monoxide.

Fluorescein isothiocyanate (FITC): is a fluorescent green dye or Fluorophore that can be tagged with cells, amino acids, and other macromolecules to be visualised in advanced imaging techniques like flow cytometry.

Fluorescence microscopy: Is an optical microscopy uses fluorescence principle to illuminate and image live and dead cells. The fluorophore like FITC tagged with the sample, absorbs light and re-emits it at a longer wavelength (**fluorescence**) which is collected and turned into an image

Fluoridation: is the technique of adding fluoride into drinking water to prevent tooth decay. Fluoride compounds, such as sodium fluoride or sodium fluorosilicate, can bind to the outer layer of the tooth, making it harder and preventing acid eruption caused by bacteria.

Fly ash: also known as coal ash, is a fine, light-weight particle that is produced during the coal combustion. As fly ash resembles particle matter (PM), it can induce lung disorders such as asthma. Fly ash is used in bricks and other building products as an alternative to cement.

Food chain: it is the sequence of food resource system or the energy flow from lower tropic organisms like plants to higher tropic organisms like goats, rabbits, and humans.

Food web: it is the interconnection of multiple food chains. For example, Plants can be eaten by rabbits and goats, while rabbits can be eaten by foxes and humans. As a result, organisms such as plants and rabbits are considered common prey, and there may be multiple predators.

Fossil fuel: is a non-renewable source of energy derived from materials found in the Earth's crust, such as coal, oil and natural gas which are formed over a time by dead and decomposing organisms.

Fourier transform infrared (FTIR) spectroscopy: is an analytical technique that uses infrared radiation to provide information on the functional groups present in a molecule. The interferogram in the spectroscopy collects the incident IR rays from the molecular vibrations and converts it into IR absorption (transmittance) and emission spectra between the range of 400-4000 cm^1

Frankel defect: is a vancant point or void that develops when atoms or ions move from their original location in the crystal lattice and occupy an alternate interstitial region, causing a defect in the structure of the crystal lattice.

Fuel cell: is an electrochemical device that uses electrodes like anode and cathode to generate chemical reactions, such as oxidation and reduction, to produce electricity.

Fullerene: also known as backminsterfullerence or bucky balls, is a type of graphene sheet that has been formed into a ball-shaped hexagonal carbon structure or cylindrical carbon nanotubes (for additional information on bucky balls, see the glossary of B).

Functionalization: is the process of altering the surface chemistry of material using various functional groups such as carboxyl, hydroxyl, amine groups, and so on. The process of surface functionalizing Nanomaterials improves their stability, drug-holding capability, biocompatibility, and makes them more suitable for biomedical applications and environmental applications.

G

Gallium (Ga): It is a rare earth metal (REM) found in the minerals like gallite and bauxite. Ga is a solid metal with a low melting point that, above 30 °C, turns into a liquid silvery metal substance. Gallium is used in LEDs, mobile phones and as an electrode in electrochemical applications.

Gamma rays: are short wavelength and high energy electromagnetic wave. Gamma rays can be are most invasive rays and can be used to destroy cancer cells and to break certain chemical interactions between molecules, hence they are utilised in the synthesis of nanomaterials such as platinum.

Gas Condensation: Condensation is a natural phenomenon in which the gas phase changes into liquid phase at a high pressure and low temperature. **Gas condensate** – Natural gas condensate or natural gas, where the hydrocarbons in liquid phase were previously present in gas phase.

Gaseous waste: Toxic gases such as carbon dioxide, carbon monoxide, sulphur dioxide, nitrous oxide, and methane are discharged into the atmosphere via industrial sources, transportation, combustion and incineration process.

GCMS: Gas Chromatography and Mass spectroscopy is an analytical technique to separate and identify gas molecules. GCMS can be used to identify the gas pollutants like methane, volatile organic compounds and other food contaminants like pesticides.

Gel electrophoresis: Gels, like Agarose is used in the gel electrophoresis process to separate biological macromolecules like DNA and RNA based on their size or molecular weight. Refer electrophoresis for more details

Gene: It is a nucleotide sequence (DNA/RNA) found on chromosome. Gene encodes for a specific protein and carries hereditary information (trait) such as a person's height and colour

Gene sequencing: It a process to identify the pattern of the base pairs (ATGC) along with the phosphate and sugar groups present in the fragment of DNA or RNA. Gene sequencing is done by using bioinformatics tools like FINDER, FragGenScan and etc.

Gene Therapy: It is a therapeutic process that modifies the defective gene or replaces healthy gene to treat various genetic diseases, or diseases like Cancer, Alzheimer and Parkinson etc.

Genetic engineering: A biotechnological tool used to modify the gene of an organism with improved traits. For example, Genetically modified plants like GM Corns were developed with larval pest resistant genes.

Genomics: It is the study of an organism's genetic information, like understanding a complete set of gene-DNA/RNA, including its structure, base pair ordering, and functions like protein coding and to identify the gene sequences responsible for particular disorders.

Geodesic domes: Buckminster Fuller created the geodesic dome, a sizable hemispherical structure made of a polyhedron with a triangle shape. The Geodesic domes served as a model for the construction of the Bucky Balls or C60 structure. Refer to buck balls for more details

Germanium (Ge): It is a greyish white metal and a semiconductor material with a band gap of 0.67 eV. GE is often used along with silicon (SiGe) as an electrode in sensing and solar cell applications

GHG: Greenhouse gases are those that absorb radiations from sun and emit back Infrared radiations to earth's surface and cause global warming. Carbon dioxide, methane and chlorofluorocarbons, are the most significant GHGs produced from Industrial, automobile, transportation and agricultural sectors.

Gibbs free energy (G): Also known as **Gibbs energy** or **enthalpy**, is the measure of the maximum potential or the work done by a thermodynamic system in a given temperature and pressure.

GIS: Geographic Information System is a computer-based technology that collects stores, manipulates and analyse the geographical data such as aerial or scanned images of plot using hardware and software tools. GIS assists in detecting soil type, groundwater level, and so on, and generates a model based on the data stored.

Glove box: It is a sealed container designed to set up a controlled environment or desired atmospheric condition with a suitable temperature and pressure for a device like gas sensor to operate.

Glutaradehyde : It is a transparent oily organic substance that is frequently employed as a fixative to cover and protect live cells such as bacterial cells during electron microscope observation.

Gold nanoparticles (AuNPs): are nano-scale gold particles with characteristics distinct from bulk gold materials. At the nanoscale range, the gold colour of the particles changes to red and they become more reactive than their bulk gold material.

GPT: Gas permeability tester, is a device that measures the permeability or flow rate of gases from plastic, rubber films based on the porosity and pressure gradient. GPT can be used to determine the quality of food and pharmaceutical packaging materials.

Grain Boundaries: It is a defect or region that separates two differently aligned crystallite or grains. In a polycrystalline material, each grain has a particular orientation and is considered as a single domain, and the area between all these domains are called as a grain boundary.

Granular Activated Carbon: Activated carbon particles that larger in size like granules and it cannot pass through a sieve with a mesh size of 50. Granular activated carbon is an adsorbent material that is used to remove heavy metals and other contaminants from the air and water.

Graphene: It is a two dimensional sheets of carbon atoms arranged in a hexagonal manner and it resembles like a honeycomb structure. The tubular form of graphene is called the carbon nanotubes (CNTs). Refer to CNT for more details

Greenhouse effect: It is a natural process in which the greenhouse gases such as carbon dioxide, methane, etc., absorb the incoming solar radiation and releases it back into the earth's atmosphere and subsequently heats it up.

Green nanoparticles: are nanoparticles synthesized by using organic materials like plants and micro-organisms by replacing toxic chemicals used as a precursor material. Therefore, it offers an alternate to chemical synthesis process.

Green plastics: Green plastics, such as cellulose polymers or biopolymers derived from plants, are plastics that are made from sustainable resources. These plastics are eco-friendly from the point of production to the point of disposal.

Green silica: It is eco-friendly silica that is produced from microbiological sources such as diatoms or through waste reduction processes such as the conversion of rice husk and fly ash into silica, therefore providing greenness to the environment. It is a replacement to chemically synthesized silica.

Grey Goo: It is a hypothetical concept of an out of control nano-robot that copies itself multiple times and consumes all life on Earth. It is like an end of the world scenario by an over powered nano machines.

Growth promoters: Are natural or synthetic compounds, such as hormones that stimulate plant growth like plant height, flowering and grain size, without altering other physiological process. For Example, Auxin and cytokinins are natural growth hormones secreted by plants.

H

Habitat: A natural environment or place where living organism relies for resources, growth, and existence. Polar bear in the arctic region and Camel in the desert region are few examples of animals that live in distinct habitats.

Half-life: It is the process by which a radioactive element is decomposed into half of its initial concentration. It is a radioactive decay indicator that aids in the safe handling of radioactive isotopes.

Hall effect: It is the process of creating voltage difference across materials like semiconductors and conductors when a magnetic field is applied to it. Hall effect sensors can be developed using this phenomenon.

Halloysite clay: It is an aluminosilicate mineral of the kaolinite group. It is commonly found in rolled tube structure with widths of 50 to 100 nm. Hence, it is a natural nanotubes material and can be used in a variety of ecofriendly agricultural and environmental applications.

Hazardous waste: Is a type of waste that is dangerous to human and environment health. Hazardous waste includes any solid, liquid, or gaseous wastes that are radioactive, corrosive, reactive, ignitable and infectious.

Heavy metal: Metals like lead, mercury, arsenic cadmium and zinc etc., are classified as heavy metals because of their high densities and specific gravity. Because of this, heavy metal pollution is persistent and harmful when it is released uncontrollably into the environment through natural and manmade sources.

Haemolysis: also known as **erythrolysis.** It is a physiological process that occurs naturally in the body that involves the rupturing (lysis) of erythrocytes, or red blood cells (RBCs). The old RBCs are removed during this process, and are replaced with new ones.

Herbicide: It is a chemical agent that kills unwanted plants or weeds that grows along with useful crops like rice, maize and sugarcane. 2, 4 Dichlorophenoxy acetic acid (2,4 D) is an example of herbicide.

Heterogeneous catalysis: It is a type of a catalyst which is in different phase from that of the reactants or the product. For example, In Habers process to produce ammonia from nitrogen and hydrogen gas, Iron (Fe) is used as a solid catalyst.

Heterotrophes: is a type of living organism that consumes other living organisms like plants and animals, as its source of energy or food. They are also called as the consumers.

Hexagonal: It is a six sided polygon with ring like structure, which is a structure of a bucky ball with six carbon atoms connecting the ring structure.

Hierarchical assembly: It is a superamolecular assembly in which a molecule, such as proteins or DNA, arranges into a basic structure, then to a secondary structure, and to a two or three dimensional structure with a definite shape. Example a smiley shaped DNA origami structure.

High energy ball milling: It is a miller or a grinder that has metal balls which colloids at a high frequency to produce fine powder. Bulk particles are grinded into nanoparticles by using high energy ball miller.

High Throughput Screening (HTS): It is a biochemical assay that rapidly screens numerous samples at a single experimental set up (ELIZA plate) to identify the target cell or macromolecules such as proteins, antibodies and enzymes.

High-area rapid printing (HARP): It is a cutting-edge 3D printing technique that produces large objects at high speed. In contrast to stereolithography, which uses an exothermic process, the Harp 3D printer uses solid-liquid slip technology to create both hard and soft objects.

Histamine: Histamine is a neurotransmitter and a hormone derived from the amino acid histidine. Histamines regulate elements such as stress and the sleep cycle, and they are also produced by mast cells and play a role in the inflammatory response.

Histopathology: It is a microscopic evaluation method for detecting disease-induced alterations or damages in tissue structure. This is the diagnosis can be used to determine the severity of the diseases and personalized treatment can be offered.

Homeostasis: The ability of the human system to be in a state of equilibrium and the internal bodily functions are maintained even though the external environment changes inevitably.

Homogeneous catalysis: The catalyst in the chemical reaction is present in the same phase as the reactants and products. For example, in the hydrolysis of methyl acetate, HCl is used as a catalyst to produce methanol and acetic acid

Hormones: Are chemical substances produced by glands or cells to regulate the action of certain bodily activities. Hormones include thyroid hormone, growth promoting hormones (GPH), oestrogen, and progesterone.

HPLC: High performance liquid chromatography (HPLC) is used to separate individual substances from a mixture of liquid samples. The liquid samples are transported (mobile phase) into the column (stationary phase), and the different components are separated by applying high pressures ranging from 50 to 300 bar.

Humidity: It is the concentration of the water vapour present in the atmosphere. Hygrometer is an instrument used in the measurement of humidity (gaseous state of water).

Hybrid nanoparticles: Organic materials like lipid, polymers or bacterial cells are combined with one or two inorganic nanoparticles like silica and iron oxide to form hybrid nanoparticles. It can be an organic-inorganic hybrid or inorganic-polymer (organic)-inorganic hybrid formations.

Hydrocarbons: They are organic molecules that exist naturally and are composed of carbon and hydrogen atoms. Hydrocarbons are the primary substance found in natural gas, coal and crude oil which are utilised in the production of fuels such as gasoline and diesel.

Hydrodynamic diameter (Dh): Hydraulic diameter or hydrodynamic radius is referred to the size of colloidal particles in suspension. The particle size analyzer (PSA) uses the Dh parameter to determine the size of colloidal nanoparticles in suspension during Brownian motion.

Hydrogel: It is a superabsorbent material made of cross linked hydrophilic polymers that imbibes water and expands to hold it in its structure in a gel form. Materials like sodium polyacrylate and sodium alginate are used as the base material.

Hydrogen (H): It is a neutrally charged atom with one proton and one electron present in the nucleus and it is the basic element present in many organic compounds.

Hydrogen bonding: It is the strongest inter and intra-molecular bonding. The H atom forms strong covalent bond with electronegative atoms such as oxygen. For example, water contains H bonding (H_2O).

Hydrophilicity: It is a substance or surface that is attracted (philic) to water molecules (hydro). Hydrophilic materials are those that can dissolve or absorb water. When a hydrophilic surface interacts with water, it achieves optimum wettability with a minimum contact angle or contact time

Hydrophobicity: it is a substance or surface that repels (phobic) water molecules (hydro).When hydrophobic compounds are combined with water, phase separation occurs. Water usually rolls off the hydrophobic surface and has the maximum contact angle or contact time.

Hydroquinone: A crystalline compound made by the reduction of benzoquinone, It is used treat pigmentation and help in skin lightening. It also treats various skin related problems like melasma, sun damage, freckles and hyperpigmentation.

Hydrosphere: It includes the area on the earth that is covered by water bodies like oceans, lakes, ponds, ground water, glaciers, and water above the earth's surface such as clouds, water vapour and moisture.

Hydroxyapetite (HA): It is a naturally occurring mineral composed of calcium phosphate. HA in with similar composition to bone and teeth, it can be used an implant in bone replacement and in dental surgery.

Hydroxyl radical (OH^-): It is a type of oxygen carrying free radicals. In human body, OH radicals are produced as a defence mechanism against infections that are infiltrating the cells. In environment, OH radicals created in the atmosphere are known as a detergent that removes contaminants.

Hysteresis loop: It is also called as a hysteresis curve, is a graphical representation of the change in the magnetic intensity of a ferromagnetic materials over a given magnetic field.

I

Ignecious rock: In Latin, the word "ignis" signifies "fire." Igneous rocks are generated by magma or lava that cools and solidifies to form into rocks. They are also known as magmatic rocks.

Iijima: Sumio Iijima, also known by his full name, is a notable Japanese physicist who invented carbon nanotubes (CNTs) in 1991 and published an article in Nature 1993 describing about the ultrafine- single walled carbon nanotube (SWCNT).

Immobilization: It is a technique of anchoring an organic molecule such as enzyme, cells, or microorganisms to an inorganic or porous material in order protect and maintain the stability of the organic material.

Immune system: Is the body's defence system that locate and attacks foreign particles. The host defence system consists of white blood cells (WBC), lymph nodes, and bone marrow, all of which work together to defend the body against disease and infection.

Immunoassay: It is an in-vitro biochemical assay used to identify the presence of a specific analyte like proteins, antibody or antigen from a given test sample, in order to diagnose diseases and infections. Refer ELIZA for more information.

Immunotherapy: is a type of bio-response therapy that functions by modifying or stimulating the immune system to locate and combat the infection or disease. Immunotherapy has become a standard procedure in cancer treatment along with chemotherapy and radiotherapy.

Impedance: It is the measure of resistance or any kind of opposition a material or a circuit can exert to the flow of electric current. Ohm (Ω) is the unit of impedance

Implants: are biocompatible or bioinert device, typically made of ceramic or aluminium are used to replace the damaged bone and deformed tissues. For example, Hydroxyapetite (HA) a substance close to bone composition is used to make bone implants.

In situ remediation: It is a remediation technique used to treat the pollutants at the site of contamination or at the point of generation from the industrial sources. For example, Permeable reactive barriers are used for groundwater remediation.

In vitro assay: It is a lab-based experiment that is carried out using test tubes, beakers, and micro-plates. To conduct this test, samples from the human body like the cells, tissues, blood, or urine are obtained, stored, and used.

In vivo assay: It is an assay conducted on a whole living organism, In pharmaceuticals, the clinical trials or testing the drug efficiency are performed using animals models like rodents, rabbits, monkeys and humans. Plant, insect and micro-organism are used as models for in vivo assays.

Incineration: It is a waste reduction process that involves burning waste materials at high temperatures and converting it to ash, flue gas, and heat. The hazardous and biomedical infectious wastes are treated by incineration process.

Infilteration: It is a penetration process in which water from the surface travels deep into the horizons of soil and refills the groundwater following precipitation.

Infrared (IR) spectroscopy: It is a spectroscopic technique that gives fingerprint information on the functional groups present in a molecule and is also known as Fourier transform infrared spectroscopy. For more information, see FTIR.

Insecticides: They are synthetic substances that are used to kill insects like cockroach, bee, beetles, bugs, caterpillar, butterfly, housefly mosquito, dragonfly, snails and so on. In agriculture insecticides like DDT are used to kill pest insects that damage the crops.

Interferometry: also referred to as interferogram, is a technique that superimposes two waves deflected in different direction into one to gather information. This technique is used in instruments like FTIR, and it is referred as the heart of the instrument.

Inter/ Intra-molecular interactions: Intermolecular interactions are those that occur between two molecules. For example, DNA and protein binding in a cell structure. Intramolecular interactions occur within the atoms in a molecule. For example, the bond between the H_2O and HCl.

Interstitial tissues: They are connecting tissues or fibrous tissues composed of elastin, collagen, and fibroblast that are found between cells and organs. Interstitial tissues are responsible for regulating hormones and immunological responses.

Interstitial defect: It is a point defect in the crystal structure that occurs when a one or more foreign electrons occupy the interstitial spaces in the crystal lattice and cause defective structure.

Ion channel: The gated ion channel, refers to the integral proteins that acts as a gate way for specific ions like Na^+ and K influx to the cell by passive diffusion process. Ion channels maintain homeostasis in a cell.

Ionic bonds: They are formed by sharing of the valence electron between molecules or by strong electrostatic bond between electronegative and positive atoms such as HCl and NaCl.

Ionising radiation: High-energy electromagnetic radiation, such as gamma rays and X-rays, can ionise molecules and cause adverse health infects like skin burns, mutations, birth defects and cancer.

Ionization: It is a process in which an atom becomes reactive and charged up either by losing or gaining an electron. This technique is used in GCMS to separate molecules into ion to the quantity the elements present.

Ionophores: are ion channel that transport certain cations like Ca^+, Na^+ and K across the phospholipid bilayer, this occurs when the protein pores are not present.

IPR: It is abbreviated as Intellectual Property rights; it is a protection of the man's novel intellectual creations such as ideas, methods, designs, inventions, products and symbols.

Iron oxide nanoparticles (Fe NPs): Fe NPs are hematite (Fe_2O_3) and magnetite (Fe_3O_4) with diameters smaller than 100 nm. These are typically synthesised by hydrothermal method using ferrous sulphate or ferric chloride as precursors. Fe nanoparticles are biocompatible and used in a variety of biomedical and environmental applications.

Irrigation: It is the process of supplying water to the agricultural crops. There are several types of irrigation like surface irrigation, drip irrigation and sprinkler irrigation.

ISO standards: It is the acronym for International Organisation for standards; it is an international non-governmental organisation that ensures the safety and the quality of process and products

Isotope: These are same chemical elements for example isotopes of carbon (C) are C^{12}, C^{14} and C^{20}, with same atomic no. 6 but different mass no. as 12, 14 and 20 and also exhibit different property. For example, C^{14} is radioactive

Itai itai disease: also known as ouch ouch disease, caused by chronic exposure to cadmium (cd) through food and water, which causes diseases like osteoporosis i.e., weak and brittle bones as well as kidney failure. The mining industry is primary source of cadmium contamination.

J

J Curve: It is a graphical representation denoting the J shaped growth or the exponential growth of a population in a favourable environment with proper nutrition

Jablonkshi diagram: It is a diagram that represents the different energy state of the electron. The ground state, the excited energy state and the transition states where the electron loses its excess energy and returns to the ground state.

JET and flash imprint lithograph (J FIL): It is a form of nano-imprint lithography that is used to create high-resolution patterns by dropping UV curable resist through inkjet (Jet spray) over a template pattern, and the resist is cured after UV flash exposure to create the desired pattern

J J coupling: It is also referred as indirect dipole-dipole interaction. J represents the total angular momentum, JJ coupling is the interaction between the angular momentum of electron and nucleon in an atom which indirectly connects the two nuclear spin orbits.

Johnson-kendal Roberts: JKR theory is an adhesion or contact theory that studies the deformation in the surface caused when two objects come into contact. It formulates the contact pressure or elastic energy and surface energy loss.

Joule (J): is the derived unit of work or energy in an internal system. It is a small amount of energy or work done by a force of one Newton (push or pull) to displace an object to one meter.

K

K selected species: They are a stable population with a higher carrying capacity (K) and have the ability to compete for resources in a specific habitat. They produce fewer offspring, but they are larger and have a longer life expectancy. Blue whales, giraffes, elephants, and humans are examples.

Kaolinite: It is a naturally occurring hydrothermally produced aluminosilicate clay mineral. It contains alumina in the octahedral layer and silica in the tetrahedral layer. Kaolinite due to its differential surface chemistry has huge industrial value.

Kelvin (K): It is the SI unit of temperature, the absolute zero temperature in Kelvin is equal to -273°C

Keratin: It is a fibrous protein produced from the epidermal cell called the keratinocytes. Keratin is the principal structural protein found in hair, skin, and nails that gives them strength.

Key stone species: Are species whose growth and survival have a significant impact on the ecosystem and establish how the ecosystem functions. For example, honey bees aid in cross-pollination, and parrotfish eat overgrown algae on coral reefs.

Kinetic energy (K E): It is the energy an object possesses during the motion. To accelerate the object an initial force is applied, and that force is transferred to the object and the object from the state of rest moves at a particular velocity.

Kjeldal method: It is a basic analytical method introduced by John Kjeldal to determine the nitrogen content present in the organic and inorganic substances.

Kyoto protocol: It is an international treaty formed by the United Nations Framework Convention on Climate Change (UNFCCC) to limit CO_2 emissions in the atmosphere in order to mitigate the effects of global warming.

L

Lab-on--chip (LOC): also called as Chip based assay, is a miniaturized device with nano and micro channels containing bio-recognition materials that can perform several assays at high resolution.

Landfill: It is site constructed to dump non-hazardous and biodegradable waste like organic waste from households and to dump rubbish. Modern landfills are also constructed to dump hazardous waste to protect it from groundwater contamination.

Langmuir-blodgett (LB): It is a method to deposit self-assembled monolayers over a surface of thin films (**LB films**) thro air-water interface. LB method is used for designing the electrodes.

Larynx: Is a hollow tube present above the throat. It also has vocal cords and is known as the voice box.

Laser method: It is a physical method of producing nanoparticles that involves the use of a laser beam to break down the precursor molecules. Laser pyrolysis is one of the gase phase laser methods used to create nanoparticles such as carbon nanotubes, silicon nanoparticles, and magnetic nanoparticles.

Lattice: A crystal lattice is a space or a point where the atoms or ions are located in a repeated order in a crystal.

LDH assay: It is a cyctotoxicity assay that quantifies the amount of lactate dehydrogenase (LDH) that is released from a damaged cell to the elements of the surrounding medium.

Leachate: It is a contaminated liquid that oozes out from a solid waste present at the disposal site like a landfill. Leachate can percolate into the soil and contaminate the groundwater.

Lead (Pb): It is a heavy metal that contaminates the soil and drinking water. The lead contamination generates through fertilizer use, smelting, pipes and paints.

LED: Light-emitting diode, when an electric field is sent through a light-emitting diode, it creates light of various colours or luminosity. The LED light operates on the electroluminescence principle.

Lentic ecosystem: It is the fresh water zone near the land surface like lake, lagoons and the ponds where the water is stagnant and is used for domestic purposes. The wetlands are also included in the lentic ecosystem.

Leukocyte: It is a white blood cell (WBC) that is made in the bone marrow and found in the lymph tissue and it is a part of the body's immune system. The primary role of leukocyte is to fight infections and disease.

Line defect: A dislocation in the arrangement of the atoms in a crystalline substance results from a slip in the direction of the crystal planes.

Limit of detection (LOD): It is a detection threshold, where the lowest concentration or the signal from the sample that can be detected at a degree of confidence or at a statistical significance.

Lipophilic: Materials with have strong affinities for lipid molecule. They are also called as hydrophobic.

Liposome: It is a minute artificial spherical sac of phospholipid molecule enclosing a water droplet to carry drugs or other substances into the tissues.

Lithium (Li): It is a silvery white alkali metal that is used in the batteries of electronic cars, laptop and phones. Li-ion batteries store more energy and make the charge/discharge process more consistent. The Li-ion batteries usage reduces the CO_2 emission and is eco-friendly batteries.

Liquid crystal display (LCD): Electronic display that functions by altering the electric current applied to a liquid crystal to obtain different optical properties. LCD is used in monitors, phones and projectors.

Liquid phase separation: When two miscible substances, such as water and oil, are combined to form an emulsion, there is a chance that the liquid will phase separate into distinct droplets over the oil

Lithography: It is a process of producing patterns on a semiconductor material to use it as an integrated circuit or sensors.

Lithosphere: It is the rigid and rocky uppermost surface of the Earth which contains the crust and upper mantle. It is the surface where terrestrial living organisms like plants and animals can be found.

Lotic ecosystem: It is an aquatic ecosystem which is characterized by the constant movement of water near the land surface like the river and the streams.

Luminescence: also known as photoluminescence, is the emission of light or photons from a substance caused by internal electronic transitions such as electron excitation and recombination. Refer electroluminescence for more details

Lycurgus cup: It is an ancient Roman glass cup that is made up of ultrathin glass-dichoric glass material that changes colour when light pass through it in different directions.

M

Macromolecule: Large molecule formed by the aggregation of many units of small molecules or it is the polymer of smaller molecules. For Example, Protein is the macromolecule formed by many units of amino acids; other examples are DNA, lipids and carbohydrates.

Magnetic resonance imaging (MRI): It is a non-invasive technique that uses strong magnetic fields and radio waves not X-rays, to scan the body organs and produce anatomical images. It is used for the detection of disease and other infection.

Magnetism: is the creation of a magnetic field when an external electric field is introduced, causing a magnet to exhibit magnetic forces like attraction or repulsion.

Mass spectrometer: It is an instrument which identifies and quantifies an ion or an atom by measuring its mass or weight. The ionizing source and detector plays an important role in the separation and the quantification of ions. It is used along with instruments like GCMS. Refer GCMS for more details.

MEMS: Micro-electro-mechanical systems, or MEMS for short, are tiny machines that integrate mechanical and electrical functions and are constructed on a chip with memory and computer code. Using this tool, microsensors, accelerometer and inkjet printers can be designed.

Mercury (Hg): It is a heavy metal and a persistent contaminant that can bioaccumulate in living organisms and cause chronic sickness. Minamata disease is caused by mercury poisoning, which damages neurological functions.

Mesoporous: Materials with 2-50 nm pore size, they are also called as super nanoporous materials. Mesoporous silica materials are used as a adsorbent, catalyst and as a inorganic component in metal-organic framework.

Mesothileoma: It is a type of a cancer that occurs in the mesothelium tissues that covers major organs like lung, abdomen, liver and heart

Mesosphere: It is one the layer in the atmosphere, which lies in middle of the stratosphere and thermosphere. It is about 30-85 km from the earths surfaced at this altitude the atmospheric temperature decrease and called as the coldest region.

Metal organic framework (MOF): It is a hybrid between inorganic material and an organic material which are porous and crystalline in nature. MOFs are used as carrier to transport drugs and nutrients and are also used as an adsorbent. Refer Hybrid nanoparticle for more details

Metamorphic rock: Metamorphorphic rocks are those that have altered composition and structure from that of existing rocks, such as igneous rock, as a result of high temperature and pressure over time.

Methane (CH_4): It is a colourless, volatile gas utilised as fuel. Like CO_2, it has a negative impact on global warming. Agriculture sector is the largest contributor to the emission of methane gas.

Micelle: They are amphiphilic molecules that like surfactants and form a spherical superamolecular structure in a colloidal suspension, with the hydrophilic head attaching to the liquid and the hydrophobic tail projecting toward the centre. Drugs that are both hydrophilic and hydrophobic can be transported using micelles.

Microfluidics: It is a developing technology that investigates the behaviour and dynamics of fluids in a micro-channel, whereas nanofluidics passes through nanoscale structures.

Milling: Milling or ball milling is a grinding machine is a physical method to cut down or break materials; it is one of the top down approach to synthesize nanomaterials from bulk materials. *Refer Ball milling for more details*

MMT: Montmorillonite (MMT) is a phyllosilicate clay mineral with nanoporous layers in its internal structure and it is also called as molecular sieves. MMT can also be size reduced to be used as an ecofriendly nanoadsorbent material.

Molecular beam epitaxy (MBE): Epitaxy denotes- "on the top assembly". MBE is a physical vapour deposition process, where atoms or molecular beam is flashed over a heated substrate to deposit nanoparticle under high vacuum condition.

Molecular recognition: It is a biological process in which two molecules interact or have an affinity for one another. This interaction can occur by chemical bonding, such as hydrogen bonds, electrostatic interaction and structure-based specificity. For example, antigen-antibody interaction and enzyme-substrate interaction.

Moore's Law: It is the concept of miniaturization, predicted by an American engineer Moore in 1960s, which is applicable to recent technology and inventions. According to this law, the transistor counts on a chip will double every two years, resulting in increased computer performance.

Mutualism: It is a symbiotic relationship between two or more species, in which each species are benefited from their interaction. For example, Honey bees and flowers interaction.

MWCNTs: MWCNTs are layered carbon nanotube materials composed of hexagonal carbon structure similar to graphene sheets. Exfoliation of MWCNTs results in single and double walled carbon nanotube architectures. MWCNTs are employed in sensor, catalytic, and solar cell applications as a conductive material.

N

Nanite: Atomic scale component of nanobot.

Nano: one billionth (1/1,000,000,000).

Nano array: Ultrasensitive array denser and compact than a microarray.

Nano assembler: A nanoscale device which combines individual molecules into the structures required.

Nano balance: Balance which even weigh viruses and other nano particles.

Nano biosensor: A device that combines a biological indicator with an electrical, mechanical, or chemical sensing system

Nano biotechnology: Biological systems learnt by application of nanotechnology and used for devices

Nano bubbles: Nano sized bubbles with or without molecules or objects.

Nano characterization: Understanding the properties of nano scale materials

Nano chips: High storage chips which will be a trending in Next generation chips

Nano clusters: Aggregates of atoms, molecules or clusters in the nanoscale.

Nano cones: Objects made with carbon and carbonaceous materials, used for liquid delivery and light transport applications.

Nano composite: A material composed of two or more substances, of which at least one has a nanoscale dimension

Nano computer: Building up computer at molecular level

Nano crystal: Crystalline structure of nanoparticles with smaller dimension in order of 100 nm

Nano crystalline materials: Amorphous phase of solids with small domains of crystallinity

Nano electromechanical Systems (NEMS): Electrical and mechanical devices at nano scale

Nano electronics: Electronics on a nanometer scale resembles semiconductor devices

Nanofabrication: Assembling structure to form nanoscale materials

Nano fiber: Nano mesh formed by electro spinning, with filament diameters of 150–200 nanometers.

Nano filters: Used for filtering very minute particles. Objects used for selective filtration of molecules depending on their sizes.

Nano-flares: Spherical nuleic acids indicates species at lower concentration

Nano fluidics: Nanoscale spaces accommodated with fluids

Nano imprinting: It is called soft lithography which involves mould based printing, and the moulds have nano scale features.

Nano indentation: Used for studying mechanical properties, stiffness and hardness of the material. Force exerted by a diamond indenter is measured as it makes an indent of nanometer dimensions in a material.

Nanolithography: Writing at nanoscale pattern

Nano machine: Device doing mechanical work at nano scale

Nano manipulation: Manipulation of items at atomic and molecular levels.

Nanomaterial: Order of material size less than 100 nm

Nano medicine: Medicine manufactured at nano level

Nanometer: A unit of length in the metric system, equal to one billionth of a meter (i.e., 10^{-9} m or one millionth of a millimeter).

Nano optics: Investigation of light-matter interaction at the nanoscale.

Nano opens and Nano pencils: Objects were drawn for nanoelectronics, utilized in dip pen lithography.

Nanoparticles: Particle of size 1-100 nm

Nano pharmaceuticals: Use of Nano scale objects for the use of delivery, transport, uptake and delivery of pharmaceuticals.

Nano pipettes: Objects similar to pipettes with which controlled delivery of chemicals, reagents and light at nanometer scale areas.

Nano pore: A small pore in an electrically insulating membrane, that can be used as a single-molecule detector.

Nano porous materials: Engineered materials with nanoscale holes, used in filters, sensors, and diffraction gratings.

Nano probe: Nano scale devices used to understand the properties of materials and systems.

Nano remediation: Remediation of environmental pollutants by utilizing nanomaterials

Nanoscale: 1 to 100 nanometers range

Nano science: The study of unique properties of matter at the nanoscale; an interdisciplinary field of science combining physics, materials science, the chemistry of complex molecules, and related disciplines.

Nano sensor: Nano scale device for sensing radiation, chemicals and biological agents at molecular level

Nano shell: A semiconductor surrounded by nanoparticle that has a metallic shell

Nanostructures: Structures made out of nanomaterials.

Nano surgery: A generic term including molecular repair and cell surgery.

Nanotube: Hollow, cylindrical structures, with a diameter usually less than 5 nanometers composed mostly of carbon.

Nano whisker: A Nano scale structure that consists of brushes attached along a common spine.

Nanowire: A wire of diameter on the order of a nanometer.

NMR (Nuclear Magnetic Resonance) spectroscopy: Utilizing strong magnetic field the structure of molecules can be determined the structure of molecules.

Non covalent interactions: It is a vander waals interaction ie., weak interactions that bind together different kinds of building blocks into supramolecular entities. Also referred to as van der Waals interactions.

Non-volatile Memory: Information can be retained constantly without electric power.

Nucleation: Phase transition happens and grows outward, nucleation in liquids are gaseous bubbles, crystals, or glassy regions.

Nano-zero-valent iron (nZVI): It is inexpensive, non-toxic and a moderate reducing reagent.

O

OLED: Organic light-emitting diodes; a special class of light-emitting diodes made of only organic molecules that emit light in response to electric current.

Oligomer: Short polymer, between monomer and polymer.

Optics: The study of the behavior and properties of light including its interactions with matter and its detection by instruments.

Optoelectronics: Electronic devices that source, detect and control light. Electrical to optical transducers or instruments that use such devices

Oxidation: Loss of electrons in a reaction and the electrons are transferred to another component of a reaction.

P

PDMS (Polydimethylsiloxane) : Silicon based organic polymer used in nano imprint and soft lithography

Perovskite: Crystal structure material with the formula ABX_3 integrate into high-efficiency solar cells, optical displays, and detectors.

Phase: A part of a sample of matter that is in contact with other parts but is separate from them.

Phase diagram: Phases in sample and borderlines between different phases depicted in diagram.

Phase transport : The movement of heat, mass, and momentum in a medium.

Photolithography: Light utilized to fetch patterns into silicon warfares and it is often used for making integrated circuits.

Photoluminescence: Light excited in a body by some form of electromagnetic radiation in the ultraviolet, visible, or infrared regions of the electromagnetic spectrum.

Piezoelectrics: Production of voltage when dielectric crystal subjected to mechanical stress or can change shape when subjected to a voltage.

Pi-stacking: DNA duplex formation gets stabilized by noncovalent interaction between two aromatic rings.

Polarization: Orientation of oscillations described in the property of waves

Plasmid: A structure in bacteria that has circular (two-dimensional) double-stranded DNA distinct from chromosomal DNA

Polymer: Naturals and synthetic materials with variety of charecteristics, in which large molecules are connected by covalent chemical bonds.

Polymer Clay: Deformable composite of polyvinyl chloride (PVC) that can be manipulated similar to a clay. This usually does not contain any actual clay.

Polymer Pen Lithography (PPL): A cantilever-free scanning probe lithography method for the high-throughput patterning of nanoscale features with controllable size.

Polymerase chain reaction: DNA molecule gets copied and has amplified complementary strands

Polymorphism: The property of a chemical substance crystallizing into two or more forms having different structures, such as diamond, graphite, and fullerenes from carbon.

Pro SNA: Protein utilized as a core substance in a type of spherical nucleic acid. A type of spherical nucleic acid that uses a protein as the core to improve delivery of protein therapeutics.

Programmable Atomic Equivalents (PAEs) : A class of DNA-functionalized spherical or anisotropic nanostructures that can be likened to conventional atoms in the sense that the nanoparticle core represents the "atom" and the DNA strands represent the "bonds" between them when such structures are assembled into colloidal crystal superlattices via programmed DNA hybridization. Spherical nucleic acids are a type of PAE.

Protein: Large organic molecules involved in all aspects of cell structure and function.

Proteomics: Characterization of proteins and construction of new proteins.

Q

Quantum confinement effect: Nanocrystals caged with atoms

Quantum Dot: A semiconductor in which the electron propagation is confined in three dimensions

Quantum tunneling: Electronic transport across a barrier without medium.

Quantum well: A thin layer of narrow band gap semiconductor sandwiched between insulators. The electrical transport is confined to the thin layer.

Quantum wire: A thin line of conducting material in which electrical transport is confined within one dimension.

Qubit: The analogue of a bit in quantum computing.

Quantum confinement: Confinement of electrons when the size of matter reduces below the exciton radius of the semiconductor. At the confined state, quantum effects are exhibited by the material.

Quantum cryptography: A cryptographic system utilising quantum properties. Unsolicited visitors alter the quantum state of the system and so are detected.

Quantum dot lasers: Lasers utilizing energy states of nanomaterials.

Quantum Hall effect: Quantised resistance observed in some semiconductors at low temperatures.

Quantum mechanics: Objects of atomic, subatomic and molecular level is described by this mehanics.

Quantum mirage: The possibility of transporting information utilizing the unique properties of quantum confined electrons.

R

Raman spectroscopy: Analysis of the intensity of Raman scattering, in which light is scattered as it passes through a material medium and suffers a change in frequency and a random alteration in phase. The resulting information is useful for determining molecular structure.

Rotaxane: A molecular machine in which a molecular ring is programmed to surround and move along a molecular axle.

RNA (ribonucleic acid): It's a linear polymer present in cytoplasm of a cell that transmits genetic information from DNA to the cytoplasm and controls certain chemical processes in the cell.

S

Scaffold: Three-dimensional biodegradable polymers engineered for cell growth.

Self-assembly: Organization of molecules by self-effect without any external stimulus.

Self-repair: Ability to heal itself without external influence.

Self-replication: An ability of an assembler to replicate itself to a definite amount within a fixed period

Scanning Electron Microscope (SEM): A microscope which uses a beam of electrons to take images of the surface of a material.

Scanning tunneling microscope: Imaging of surface to its atomic accuracy to pin molecule of a surface.

Scanning Force Microscope: A very high-resolution microscope that uses a microcantilever to scan the surface of a substrate

Second-Harmonic Generation (SHG): The reflected light gets deflected at a frequency twice that of the incident light and allows the study of surface phenomenon of various images

Self-repair: A property of a material where it can correct or fix defects within itself.

Semiconductor: A material that exhibits electrical conductivity properties between those of a conductor and an insulator.

Self-Assembly: Processes in which a disordered system of pre-existing components forms an organized structure or pattern as a consequence of specific, local interactions among the components, without external direction.

Sensor: A device that detects a change in environment or property.

Simulation: Mechanism for predicting computationally useful functional properties of systems.

Single-molecule studies: The analysis of individual molecular properties in contrast with the study of bulk properties.

Single-source precursors: In materials science, the starting materials for semiconductor devices; powdered materials with uniform chemical composition throughout the mixture.

Small-Angle X-Ray Scattering (SAXS): A high-energy X-ray-based characterization technique used to analyze the structural properties of the unit cells that make up nanocrystalline materials.

Solar Cell: Semiconductor or organic device used to harness solar energy and convert into electrical energy.

Sol-gel materials: Gels, glasses, and ceramic powders synthesized through the sol-gel process; organic-inorganic composite materials.

Sol-gel process: A chemical synthesis technique for preparing gels, glasses, and ceramic powders generally involving the use of metal alkoxides.

Solid-state reactions: Transformations that occur in and between solids other phases to produce solids.

Solution: A homogeneous mixture composed of two or more substances. In such a mixture, a "solute" is dissolved in another substance, known as a "solvent."

Spherical Nucleic Acid: Nanostructures that typically consist of a densely packed, highly oriented, radial arrangement of nucleic acids affixed at one end to a three-dimensional, spherical nanoparticle core.

Spintronics: Electronic devices exploiting the spin of electrons, in addition to charge.

SPM: Scanning probe microscope (SPM), including AFMs and STMs, in which effect of interaction of a sharp probe with the sample is measured to infer atomic structure of the material.

Substrate: A wafer that is the basis for subsequent processing operations in the fabrication of semiconductor devices.

Superconductor: An object or substance that conducts electricity with zero resistance.

Super hydrophobicity: Extreme water repellence.

Super lattice: A periodic structure comprised of nanoparticle building blocks.

Supramolecular synthesis: Non covalent interactions forms the molecular complexes.

Surface to Volume Ratio: The ratio of exposed surface area to volume of the particle.

Suspension: A heterogeneous fluid containing solid particles that are sufficiently large for sedimentation.

SWNTs: Single-walled carbon nanotubes is an allotrope of sp2 hybridized carbon atoms similar to fullerenes

Synthesis: Union of simpler compounds or elements to build a complex compound.

Synthetic methods: Techniques for the design and creation of new materials in the laboratory.

T

Template synthesis: The engineered design and creation of materials of controlled size, shape, and surface chemistry.

Template: In cell and molecular biology, the macromolecular model for the synthesis of another macromolecule.

Top-down assembly: The process of removing material from a larger structure to create smaller ones.

Toxicity: The extent to which a chemical substance is poisonous or, through chemical action, destroys living tissue.

Toxicokinetics: Toxicokinetics is the study of how potentially toxic substances can enter the body and how they are dispersed when taken in by an organism

Transistor: Regulation of flow of current between two electrodes of solid state electronic device based on presence of semiconductor material.

Transmission Electron Microscopy (TEM): The use of electron high-energy beams to achieve magnification close to atomic observation.

U

Ultra-Violet light: Electromagnetic radiation with a wavelength shorter than that of visible light, but longer than x-rays, in the range 10nm to 400nm, and energies from 3 eV to 124 eV

UV/VIS (Ultraviolet-Visible) Spectroscopy/Spectrophotometry: Determination of concentrations of an absorbing species in solution. This technique uses light in the visible and adjacent near ultraviolet (UV) and near infrared (NIR) ranges to achieve this quantitative analysis.

V

Vapor deposition: Thin films of one substance onto a surface of another substance and this process commonly used in semiconductor industry.

Virtual reality system: Illusion of being in three dimensional world interconnected by combination of computer and interface designs.

Virus: Cell invaded by a parasite that acts malicious through execution.

Vesicles: It store, transport or digest cellular products and wastes, a relatively small and enclosed compartment, separated from the cytosol by at least one lipid bilayer.

W

Watson-Crick pairing: The hydrogen bonding motifs behind the programmable specificity of DNA-DNA interactions, where nucleobases will preferentially bind with their complements (i.e., adenine pairs with thymine and cytosine pairs with guanine).

Weak Link Approach (WLA): A supramolecular coordination-based chemistry approach, where an effector molecule is used to selectively and reversibly break a weak metal-ligand bond in a large molecular structure, opening a nanoscale chemical pocket.

Wet nanotechnology: Bio nanotechnology, where Nano systems function in the presence of water.

Wetting: Wetting occurs if a liquid is in contact with another phase, typically a solid substrate, with the substrate exerting an attractive force on the liquid molecules.

X

X-ray analysis: Radiation of X-ray radiation to identify elemental composition and to identify crystal structure by diffraction patterns

X-ray diffraction: The scattering of x-rays from a crystal, resulting in an interference pattern used to determine the structure of the crystal.

Y

Young's modulus (polymers): Ease of deformation in property of a material

Ytterbium: A rare-earth metal and member of the lanthanide family on the periodic table. Commonly found in Nano medicines, up conversion nanoparticles etc.

Z

Zeta Potential: The electrical potential difference at the interface between a solid particle and liquid that it is immersed in.

❑❑❑

3

Instrumentation – Nanotechnology

Sl. No.	Name of the Instrument	Purpose /use	Types of sample
1.	AFM Raman Systems	Provides a variety of nanometric characterizations such as topography, conductivity and thermal measurements.	Polymers, Ceramics, Composites, Glass, and Biological Samples.
2.	Atom Probes	Analysis of the chemical composition of a wide range of alloys	Solid samples (Contaminated Soils)
3.	Atomic Absorption Spectrometers	Detects elements in liquid samples through the application of characteristic wavelengths of electromagnetic radiation from a light source.	Liquid samples (Beverages, water, waste water)
4.	Atomic Emission Spectrometers	To determine and quantify the elemental composition of a material.	Analysis of trace elements in soils, water, metals, biological specimens, clinical specimens, food, physical evidence such as glass, and other solids.
5.	Atomic Force Microscopes	To image the topography of soft biological materials in their native environments	Polymers, ceramics, composites, glass, and biological samples.
6.	Atomic Layer Deposition (ALD) Systems	Allows for atomic layer thickness resolution, excellent conformity of high aspect ratio surfaces, and pinhole-free layers.	Wide variety of materials (Oxides, Fluorides, metals)
7.	Ball Miller	Commonly used for crushing and grinding the materials into an extremely fine form	Solid samples (Soil, Dry plants)

8.	Brunauer- Emmeett-Teller	Used to measure the surface area of solid or porous materials.	Solid samples (Soil, Charcoal)
9.	Calorimeter	To measure the volume and heat produced during a certain time interval	Liquid, non-volatile samples.(water)
10.	Chemical Vapour Deposition	To make 2D nanomaterials and thin films on solid substrates	Solid samples
11.	Cryogenic Probe Stations	Enables cost effective, stable, reliable and convenient low temperature semiconductor testing and even cryogenic wafer-scale testing of devices and circuits.	Liquid or gaseous samples
12.	Cryogenic Systems	To cool the superconducting magnets.	Biological Samples (Plant samples)
13.	Cryo-Transmission Electron Microscopy	Analyse the biologically originated samples at molecular level.	Solid samples with low temperature (Biological Samples)
14.	Dispersion Systems	Distributed particles of one material are dispersed in a continuous phase of another material.	Liquid samples (Emulsion, water)
15.	Dynamic Light Scattering Instruments	Measures the size of particles sus-pended in liquid by measuring temporal fluctuations in the intensity of scattered light that reflects the diffusion of the particles.	Solvents, Liquid samples
16.	Electrical Conductivity Measurement Systems	To measure the conductivity of metals on-site and in the laboratory.	Solid samples (Soil)
17.	Electron Backscatter Diffraction (EBSD) Systems	Can determine the local crystal structure and crystal orientation at the surface of a specimen.	Solid samples (Plastic)
18.	Energy Dispersive X-Ray Spectroscopy	To analyze the elemental composition of solid surfaces	Solid samples (Steel)

19.	Fluorescence Microscopes	To observe the localization of molecules within cells, and of cells within tissues.	Living Organisms.
20.	FT-IR and FT-NIR Spectrometers	To quickly and definitively identify compounds such as compounded plastics, blends, fillers, paints rubbers, coatings, resins, and adhesives.	Solid samples (soil, powders products) and Liquid samples (oils, water)
21.	Gas Chromatography	It is an analytical technique used to separate and detect the chemical components of a sample mixture to determine their presence or absence and/or quantities.	Food contaminant, Environmental pollutants
22.	Gas Chromatography Mass Spectrometry	Applications of GC-MS include drug detection, fire investigation, environmental analysis, explosives investigation, and identification of unknown samples, including that of material samples obtained from planet Mars during probe missions.	Pesticides
23.	Glove boxes	Allow materials to be manipulated without risk of exposure, possibly in a pressurized or vacuum environment.	Solid and Biological Samples (plant, live specimens)
24.	High-Pressure Homogenizer	Creating emulsions and for cell analysis when relatively large volumes are being processed	Liquid samples (Emulsion)
25.	High Resolution Transmission Electron Microscopy	Provides images of atomic structure of samples	Water sample, soil, plant samples
26.	Inductively Coupled Plasma Mass Spectrometry	Measure and identify elements within a sample matrix based on the ionization of the elements within the sample	Soil samples, water samples
27.	Induction Heating	To bond, harden or soften metals or other conductive materials.	Graphite , gold, copper, steel

28.	Infrared (IR) Spectroscopy	To determine the functional groups of molecules, relevant to both organic and inorganic chemistry.	Solid, liquid or gas samples (soil, water samples)
29.	Liposome Extruder	To achieve a specified vesicle size in a reproducible and scalable process.	Fluid samples
30.	Liquid Chromatography - Tandem Mass Spectrometry	Mass based detection of targeted compounds and physical separation of the same.	Various Solutions
31.	Micro Hardness Testers	Determining a material's hardness or resistance to deformation when test samples are not suitable for macro-hardness. / To measure hardness in very local areas.	Solid samples (Metals, composites, ceramics)
32.	Microfluidic Devices	An instrument that uses very small amounts of fluid on a microchip to do certain laboratory tests	Body fluids, solutions
33.	Microinjection Systems	Microinjection processes use either metal microinjection needles or glass micropipettes to inject small liquid volumes	Fluid samples, liquid samples
34.	Micromechanical Testing	It is used to test the tensile, compression, cycling, creep and other such properties of materials	Micro solid samples
35.	Micro positioning Staging & Systems	Micropositioning systems are motor-driven stages with travel ranges of a few millimeters up to a few hundred millimeters	Solid and Liquid samples
36.	Mills for Reducing Particle Size	Ball mill is used for size reduction process on materials such as coal, pigments and clay.	Solid samples (coal,clay)
37.	Moisture Analyzers	A device that determines the moisture content with the loss on drying method and consists of a weighing and halogen heating unit	Solid, liquid and Gas samples

38.	Multi-Angle Light Scattering Instruments	Form of static light scattering detector which allows the absolute molecular weight (Mw) and potentially the radius of gyration (Rg) of a sample to be measured.	Liquid samples
39.	Nano 3D Printing	Capable of micron-resolution printing, with the capability to print thousands of parts in a single build	Solid samples (Plastic)
40.	Nano fiber production equipment	Nano fiber production equipment is designed for production of fibers with diameter in submicron range (less than 500 nm), mainly in the form of a nanofiber web supported by a substrate material.	Liquid samples (polymer, solvents)
41.	Nano imprint Lithography Equipment	Simple mechanical lithography technique involving a stamp or a template, pressed against a deformable imprint resist layer deposited on a substrate to make the contours of the template.	Solid samples
42.	Nano indentation Testers	Measuring hardness, elastic modulus, creep, and other surface properties from the nanometer to the micrometer scale	Nano scale to micro scale samples
43.	Nanolithography systems	Fabrication of leading-edge semiconductor integrated circuits or nanoelectromechanical systems (NEMS).	Solid samples
44.	Nanoparticle characterization systems	Study the features such as size, structure, composition, and many other properties of the nanoparticles, also study the novel properties such as small size, ease of suspension in liquids, and deep access into smaller units such as cells, and their structure, high surface area, and enhanced chemical, biological, and physical properties.	Solid, Liquid and biological samples

45.	Nanoparticle production systems	Generate nanoscale particles in a precisely-defined, limited range of sizes for a range of uses, including the production of formulated coatings and other composite materials.	Solid and Liquid samples
46.	Nuclear Magnetic Resonance (NMR) Spectrometers	Allows the molecular structure of a material to be analyzed by observing and measuring the interaction of nuclear spins when placed in a powerful magnetic field.	Liquid samples, Solid samples
47.	Optical Characterization Systems	Using photons of light to determine materials properties	liquid samples
48.	Optical Non-contact Profilers	Interference microscopes used to accurately measure the three-dimensional height variations on surfaces of a given test material.	3 dimensional liquid and solid samples
49.	Optical Tweezers	Used to apply precise and very localized optical forces to microscopic particles	Biological samples (living organisms)
50.	Particle Size Analyzers	Analytical technique by which the distribution of sizes in a sample of solid or liquid particulate material	liquid samples
51.	Photoluminescence Mapper	Utilizes the properties of photoluminescence to map defects in a semiconductor layer	Solid and Liquid samples
52.	Piezo Actuators	Converts an electrical signal into a precisely controlled physical displacement (stroke)	Ceramics
53.	Plasma cleansing system	Used for a variety of surface cleaning purposes before processing. It is efficient for removing surface oxidation and clearing mineral residue from surfaces.	Gaseous samples (inert gases)
54.	Pore size analyser	It measure the porosity and pore size by quantifying adsorption isotherms	Solid samples (powder, soil)

55.	*Profilometer*	Measuring instrument used to measure a surface's profile, in order to quantify its roughness	Solid samples
56.	Raman Spectrometers	Non-destructive chemical analysis technique which provides detailed information about chemical structure, phase and polymorphy, crystallinity and molecular interactions	Solid samples (powder samples), Liquid samples
57.	Safety cabinet	Protect personnel against biohazardous or infectious agents and to help maintain quality control of the material being worked with as it filters both the inflow and exhaust air.	Solid, Liquid and Biological Samples. (plant, live specimens)
58.	Sample preparation equipment	Equipment used for the preparation of physical specimens for subsequent microscopy or related disciplines - including failure analysis and quality control.	Solid, Liquid and Biological Samples
59.	Scanning Electron Microscopes (SEM)	Type of electron microscope that produces images of a sample by scanning the surface with a focused beam of electrons.	Solid sample, biological sample (leaves, seed, root)
60.	Scanning Near-Field Optical Microscopes (SNOM)	Microscopy technique for nanostructure investigation that breaks the far field resolution limit by exploiting the properties of evanescent waves.	Biological samples in Liquid form
61.	A scanning probe microscope (SPM)	An instrument used for studying surfaces at the nanoscale level. SPMs form images of surfaces using a physical probe that touches the surface of a sample to scan the surface and collect data, typically obtained as a two-dimensional grid of data points and displayed as a computer image.	Solid samples

62.	Scanning Tunneling Microscopy	It allows researchers to map a conductive sample's surface atom by atom with ultra-high resolution, without the use of electron beams or light, and has revealed insights into matter at the atomic level	Conductor and Semi-Conductor samples
63.	Scratch testers	Used to characterize film-substrate systems and to quantify parameters such as adhesive strength and friction force by using a variety of complementary methods	Solid samples
64.	SEM Raman Systems	Use scanning electron microscopy (SEM) and Raman spectroscopy to determine the nanoscale structure and chemical makeup of test samples	Solid and Liquid samples
65.	Separation Membranes	Membrane acts as a selective barrier allowing relatively free passage of one component while retaining another	Solid samples
66.	Spectrofluorometers	Instrument which takes advantage of fluorescent properties of some compounds in order to provide information regarding their concentration and chemical environment in a sample.	Liquid samples (waste water)
67.	Spectrometers	A scientific instrument used to separate and measure spectral components of a physical phenomenon.	Liquid samples (water sample)
68.	Spectrophotometers	Quantitative measurement of the reflection or transmission properties of a material as a function of wavelength	Liquid and Solid samples (soil, water sample)

69.	Spectroscopic ellipsometer	Non-destructive, noncontact, and non-invasive optical technique which is based on the change in the polarization state of light as it is reflected obliquely from a thin film sample	Multilayer semiconductor structures
70.	Spectroscopy Accessories	Determine the composition of a material sample, spectroscopy equipment analyses light that is released, absorbed or dispersed	
71.	Spray dryer	Applications of spray-drying in food processing include conversion of fruit and vegetable juices into instant powders and mixes, preparation of instant coffees and teas, drying of eggs and dairy products, ice cream mixes, and encapsulated flavors and bioactive nutraceuticals.	Liquid (Milk, fruit juices)
72.	Sputtering Systems	Plasma based deposition process in which energetic ions are accelerated towards a target	Biological samples (leaf, seed)
73.	Surface analyser	Technique for discovering the chemical structure of an extremely shallow and thin area called the surface number atomic layer of the solid matter.	Biological samples (Plant sample)
74.	Transmission electron microscopy (TEM)	Microscopy technique in which a beam of electrons is transmitted through a specimen to form an image.	Water sample, soil, plant samples
75.	Thermal analysis equipment	Instruments that analyze materials as they change with temperature relating to heat, such as freezing and boiling temperatures, the heat of fusion, the heat of vaporization	Solid samples (polymer)

76.	Thermal Desorption Equipment	Removes organic contaminants from soil, sludge or sediment by heating them in a machine called a "thermal desorber" to evaporate the contaminants.	Solid samples (soil sample)
77.	Thin Film and Coating Thickness Measurement Tools	Used to determine film thickness and refractive index of polymers, semiconductors, thin film metals	Solid samples (coated), polymer, semiconductor.
78.	Thin-Film Deposition Systems	Technique for depositing a thin film of material onto a substrate or onto previously deposited layers.	Solid samples (Polymer)
79.	Tribometers - Friction and Wear Testers	Device which is used to simulate friction and wear at the interface between surfaces in a relative motion under controlled conditions	Lubricant samples
80.	Ultraviolet–visible spectroscopy	Bacterial culturing, drug identification and nucleic acid purity checks and quantitation, to quality control in the beverage industry and chemical research.	Liquid samples (water samples)
81.	Viscometer	Viscometers measure the viscosity and flow properties of a fluid	Liquid samples and Solution
82.	Wafer Bonders	Process by which two mirror-polished wafers of any materials adhere to each other at room temperature without the application of any macroscopic gluing layer or external force.	Liquid samples
83.	X-Ray Detectors	Devices used to measure the flux, spatial distribution, spectrum, and/or other properties of X-rays.	Solid samples (Metals)

84.	X-Ray Diffractometers	Device for analyzing and measuring the structure of materials	Solid samples (carbon fibers, graphite)
85.	X-Ray Fluorescence Analyzers	Non-destructive analytical technique used to determine the elemental composition of materials	Biological samples (Plant samples)
86.	X-Ray Imaging	Understanding of characteristics of nano-materials and nanostructure of crystalline materials	Human body
87.	X-Ray Microscopy	Uses electromagnetic radiation in the soft X-ray band to produce magnified images of objects	Biological sample
88.	X-Ray Photoelectron Spectrometers	Surface sensitive, non-destructive technique used routinely to analyze the outermost ~10 nm (~30 atomic layers) of natural and engineered materials.	Solid samples (Polymer)
89.	Zeta Potential Analyzers	Does the measurements and calculations to ascertain the zeta potential of a given material	Liquid sample (Emulations)

4

Major Instrumentation – Schematic Flow Diagram

Atomic Absorption Spectroscopy (AAS)

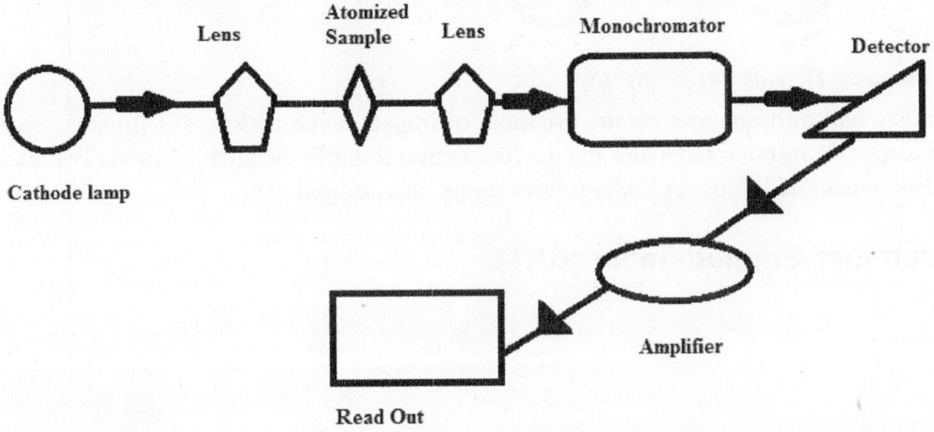

Atomic absorption spectroscopy (AAS) works based on the principle of absorption of light by free metallic ions. The quantitative determination of chemical elements through absorption of light by free atoms.

Ball Mill

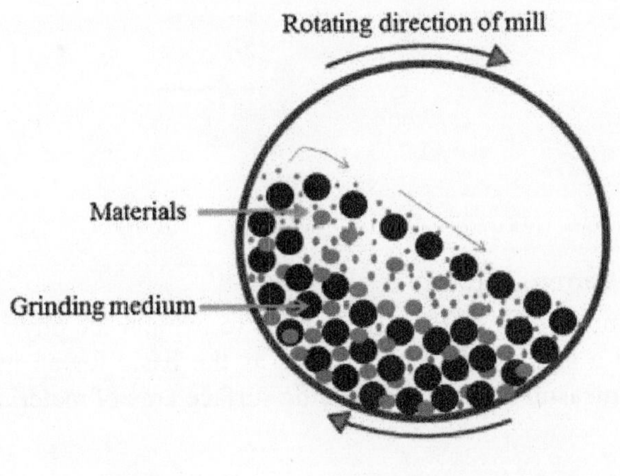

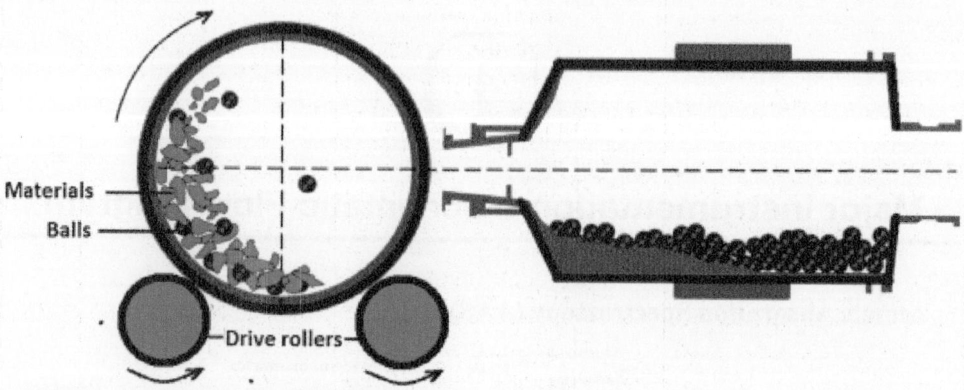

Source: (Khadka *et al,* 2014).

The ball mill operates on the principle of impact and attrition. The impact occurs between fast moving balls and the powder material while the attrition mode comes in action when the hollow cylinder rotates on its longitudinal axis.

Brunauer-Emmett-Teller (BET)

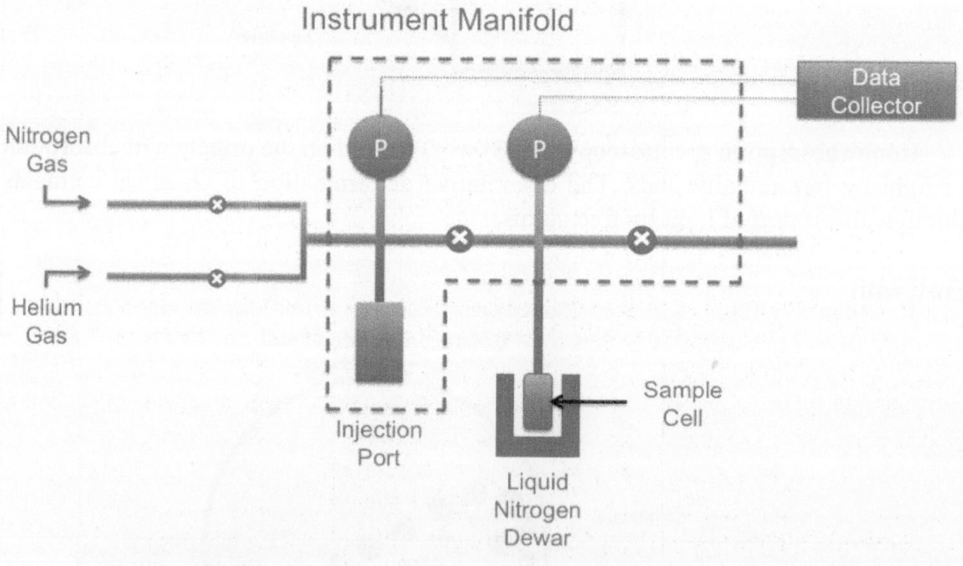

Source: (Barron, 2015).

Brunauer-Emmett-Teller (BET) theory aims to explain the physical adsorption of gas molecules on a solid surface and serves as the basis for an important analysis technique for the measurement of the specific surface area of materials.

Electro-Spinning

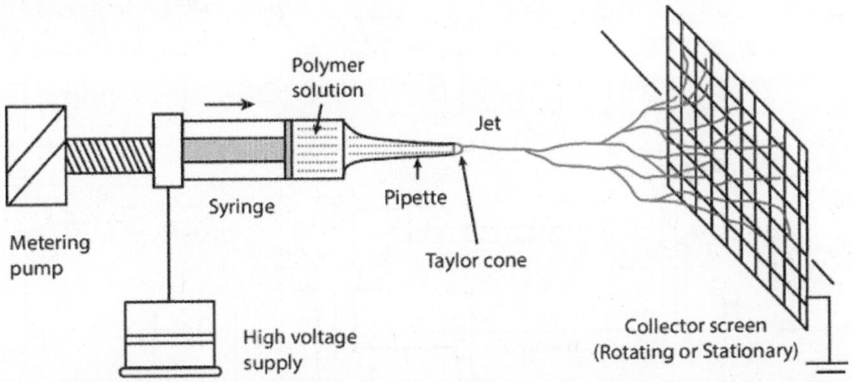

Electro spinning generate ultrathin fibres. We then discuss its renaissance over the past two decades as a powerful technology for the production of Nano fibers with diversified compositions, structures, and properties.

Field Emission Scanning Electron Microscope

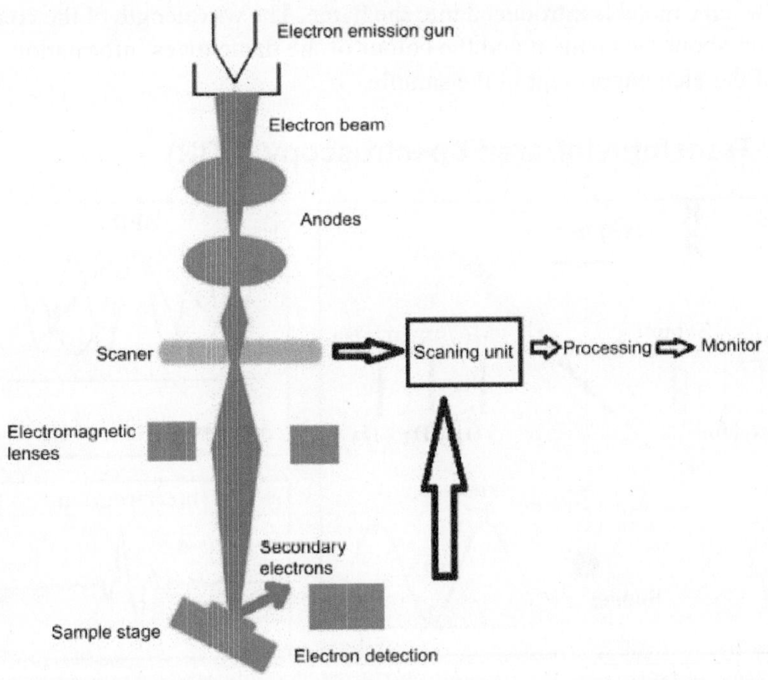

Source: (Semnani, 2017).

The FESEM is one microscope that works with electrons with a negative charge instead of light. These electrons are liberated by a field emission source. The object is scanned by electrons according to a zig-zag pattern.

Flame Photometer

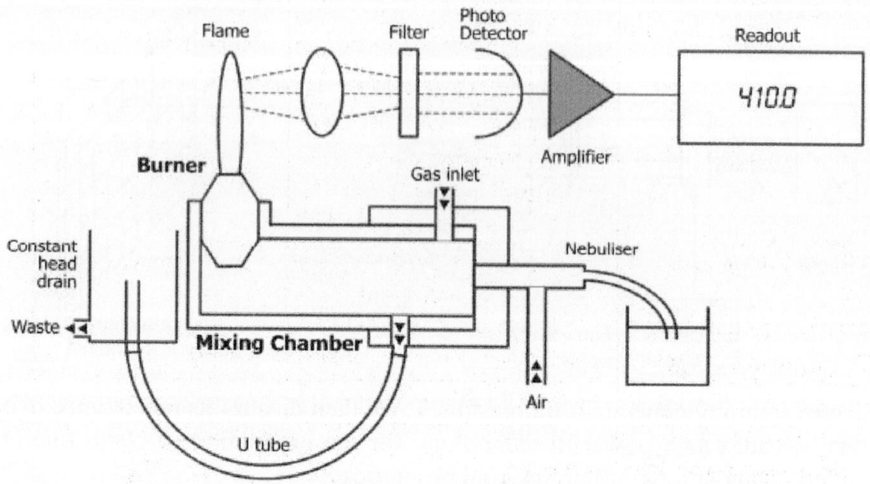

Source: (Kealey and Haines, 2002)

The principle of flame photometer is based on the measurement of the emitted light intensity when a metal is introduced into the flame. The wavelength of the colour gives information about the element and the colour of the flame gives information about the amount of the element present in the sample.

Fourier-Transform Infrared Spectroscopy (FTIR)

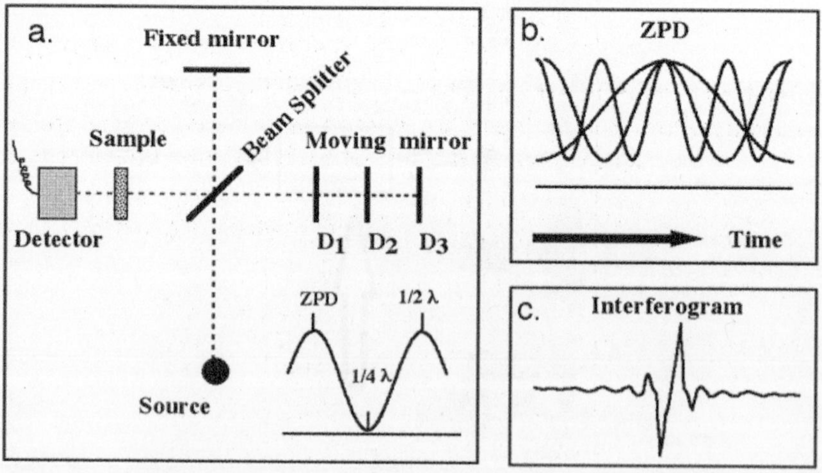

Source: (Schroeder, 2002).

When infrared (IR) radiation passes through a sample, some of the radiation is absorbed. The radiation that passes through the sample is recorded. Because different molecules with their different structures produce different spectra, the spectra can be used to identify and distinguish among molecules.

Gas Chromatography (GC)

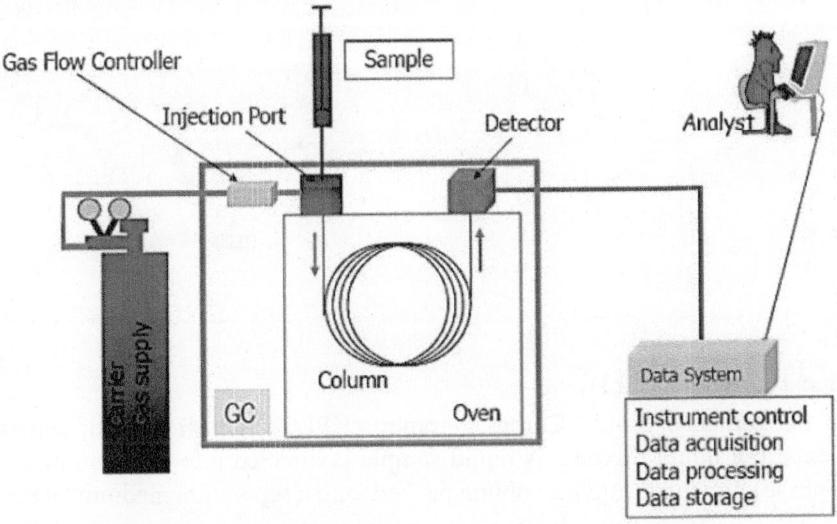

Source: (Al-Bukhaiti *et al.*, 2017).

Gas Chromatography is an invaluable tool used for separation of compounds in a mixture followed by their qualitative identification and quantification

Gas Chromatography–Mass Spectrometry (GC - MS)

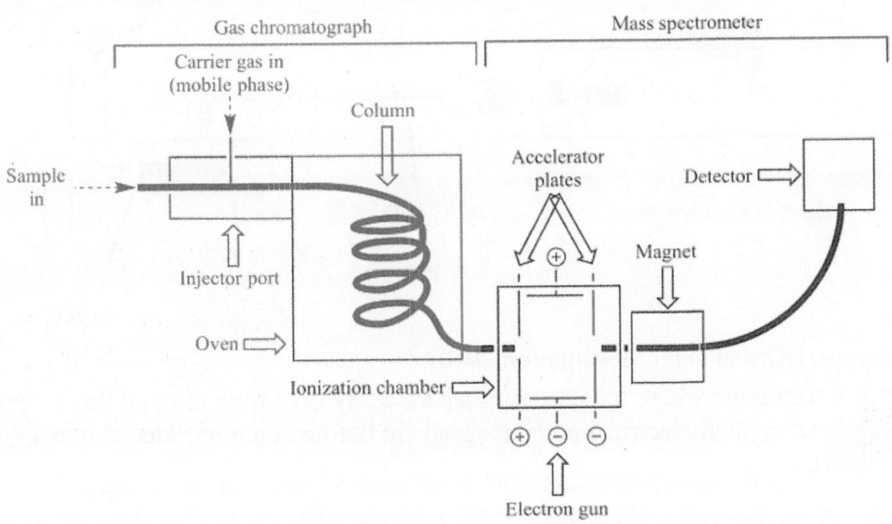

A technique to analyse the components and determine the quantity and identity of components in an analyte. The GC is fed with the analyte and the output from the gas chromatograph fed into a mass spectrometer (which determines the molecular formula of each component).

High Performance Liquid Chromatography (HPLC)

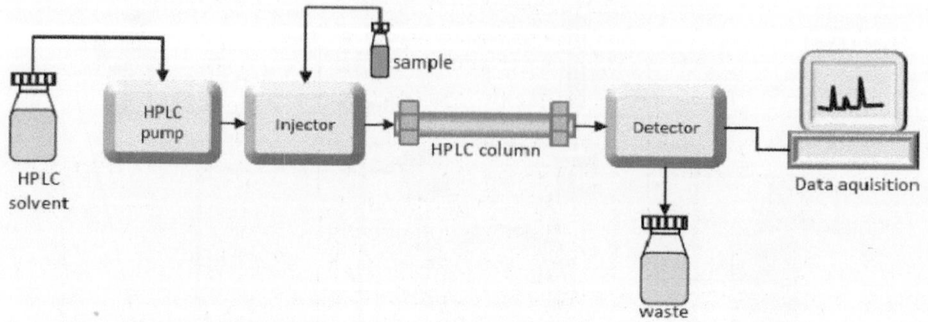

Source: (Czaplicki, 2013).

High Performance Liquid Chromatography (HPLC) is a process of separating components in a liquid mixture. A liquid sample is injected into a stream of solvent (mobile phase) flowing through a column packed with a separation medium (stationary phase).

Inductively Coupled Plasma Mass Spectrometry (ICP-MS)

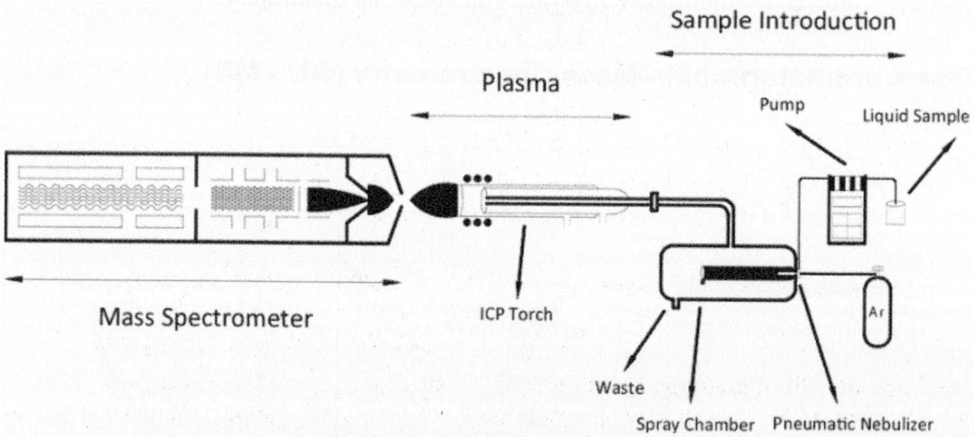

Source: (Kashani and Mostaghimi, 2010).

It is a technique where ions are first separated by drift time through some neutral gas under an applied electrical potential gradient before being introduced into a mass spectrometer.

Nuclear Magnetic Resonance Spectroscopy

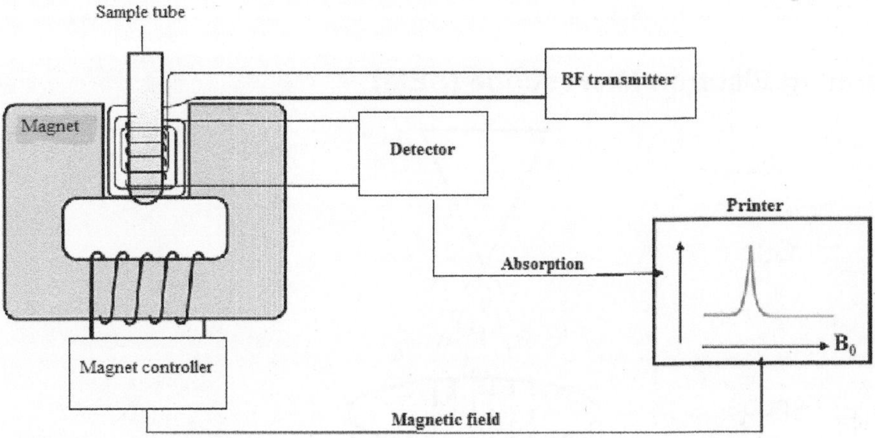

Source: (Zia *et al*, 2019).

Nuclear magnetic resonance spectroscopy is a spectroscopic technique to observe local magnetic fields around atomic nuclei. The sample is placed in a magnetic field and the NMR signal is produced by excitation of the nuclei sample with radio waves into nuclear magnetic resonance, which is detected with sensitive radio receivers. The intra molecular magnetic field around an atom in a molecule changes the resonance frequency, thus giving access to details of the electronic structure of a molecule and its individual functional groups.

Particle Size Measurement

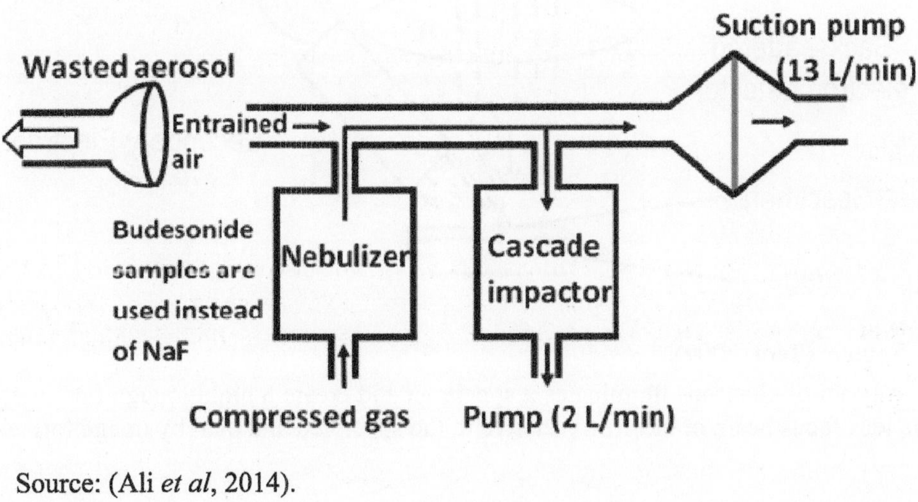

Source: (Ali *et al*, 2014).

Particle size analysis is used to characterise the size distribution of particles in a given sample. Particle size analysis can be applied to solid materials, suspensions, emulsions and even aerosols.

Scanning Electron Microscope (SEM)

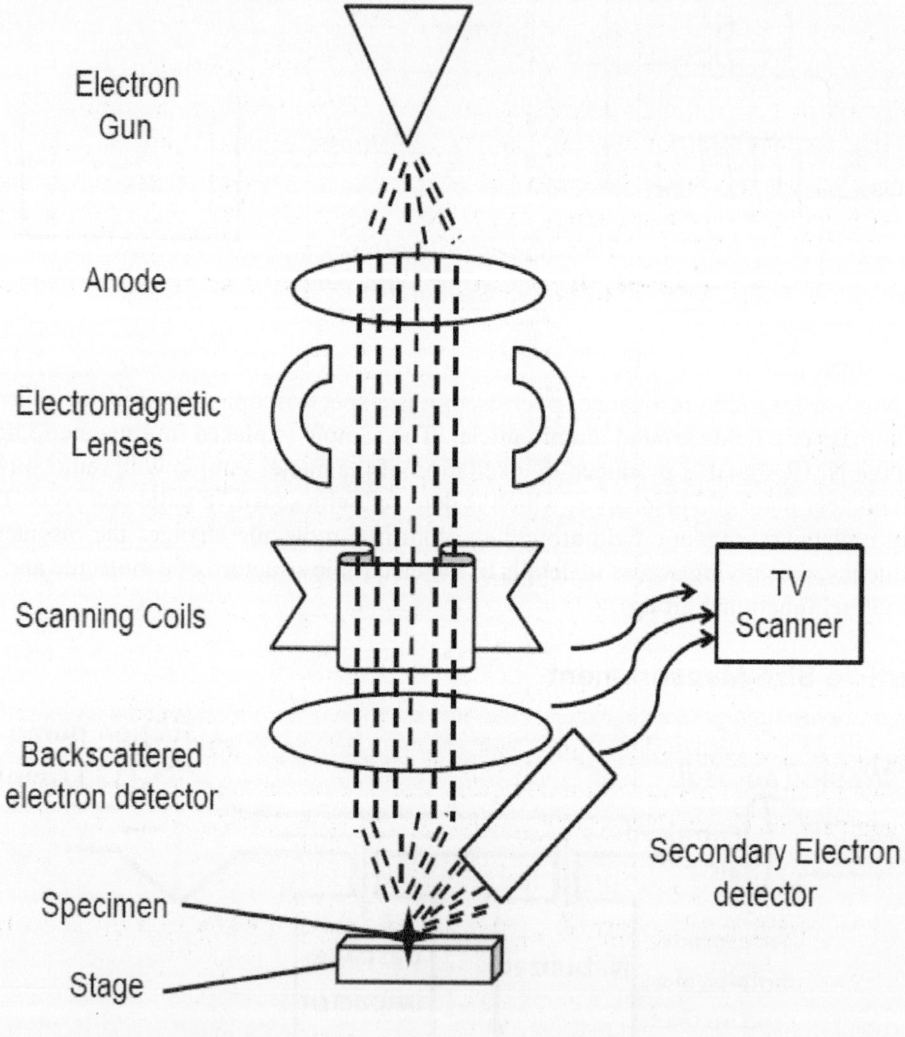

Source: (Tare, 2009).

A beam of electrons illuminates a specimen and create a highly magnified image. The lens focus beam of electrons relative to the specimen and thereby image formed.

Transmission Electron Microscopy (TEM)

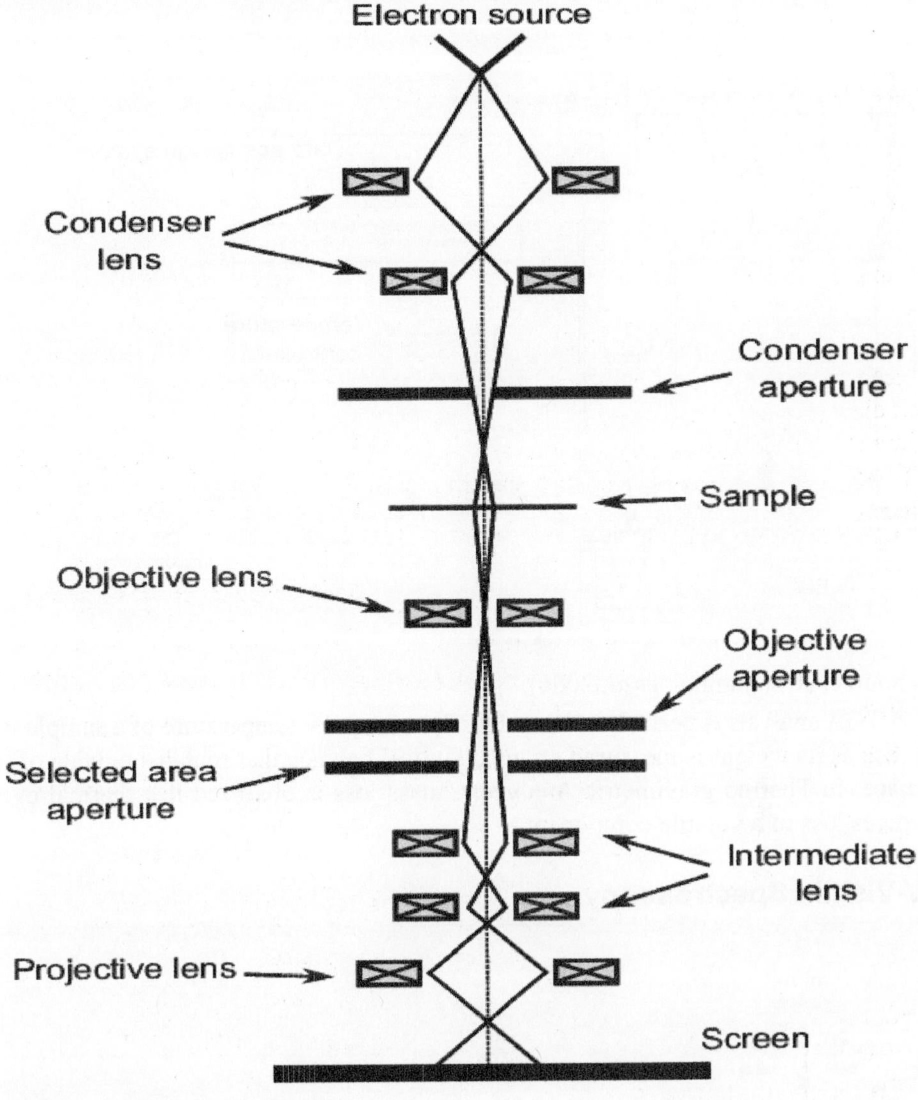

Sourcc: (Marturi, 2013).

The specimen is mostly ultrathin section and the beam of electrons transmitted through a specimen to form an image.

Thermo gravimetric Analyzers

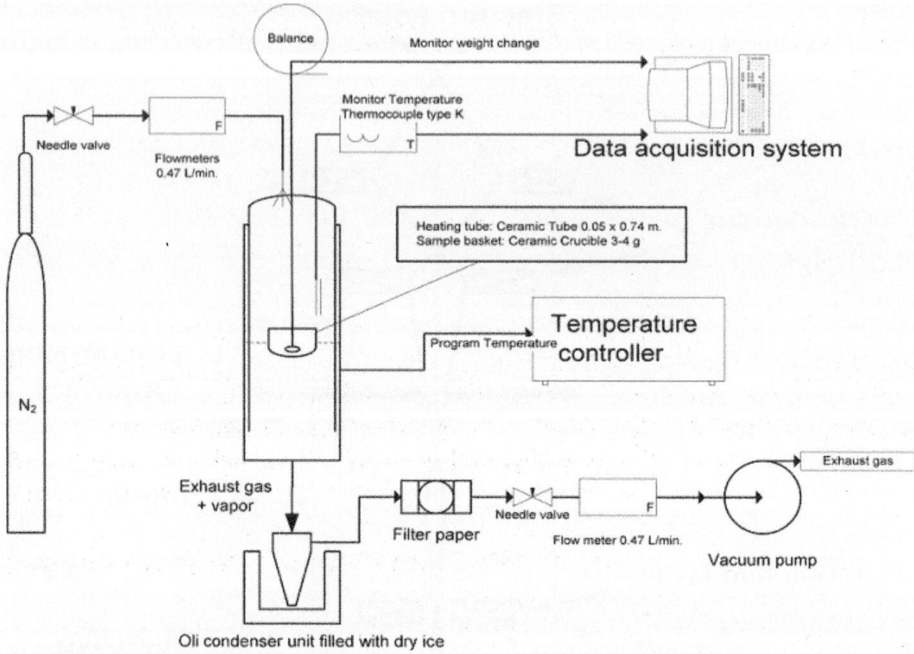

Source: (Unapumnuk *et al*,.2006).

A TGA analysis is performed by gradually raising the temperature of a sample in a furnace as its weight is measured on an analytical balance that remains outside of the furnace. In Thermo gravimetric Analyzers, mass loss is observed if a thermal event involves loss of a volatile component.

UV-Visible Spectroscopy

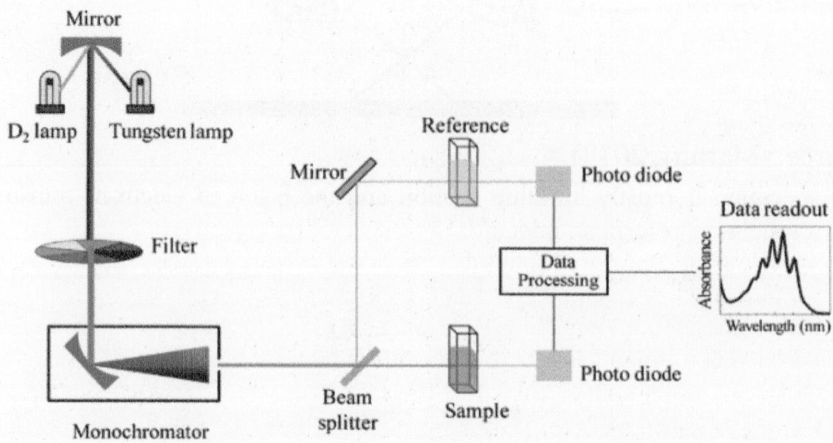

The Principle of UV-Visible Spectroscopy is based on the absorption of ultraviolet light or visible light by chemical compounds, which results in the production of distinct spectra. Spectroscopy is based on the interaction between light and matter.

X-Ray Photoelectron Spectrometer

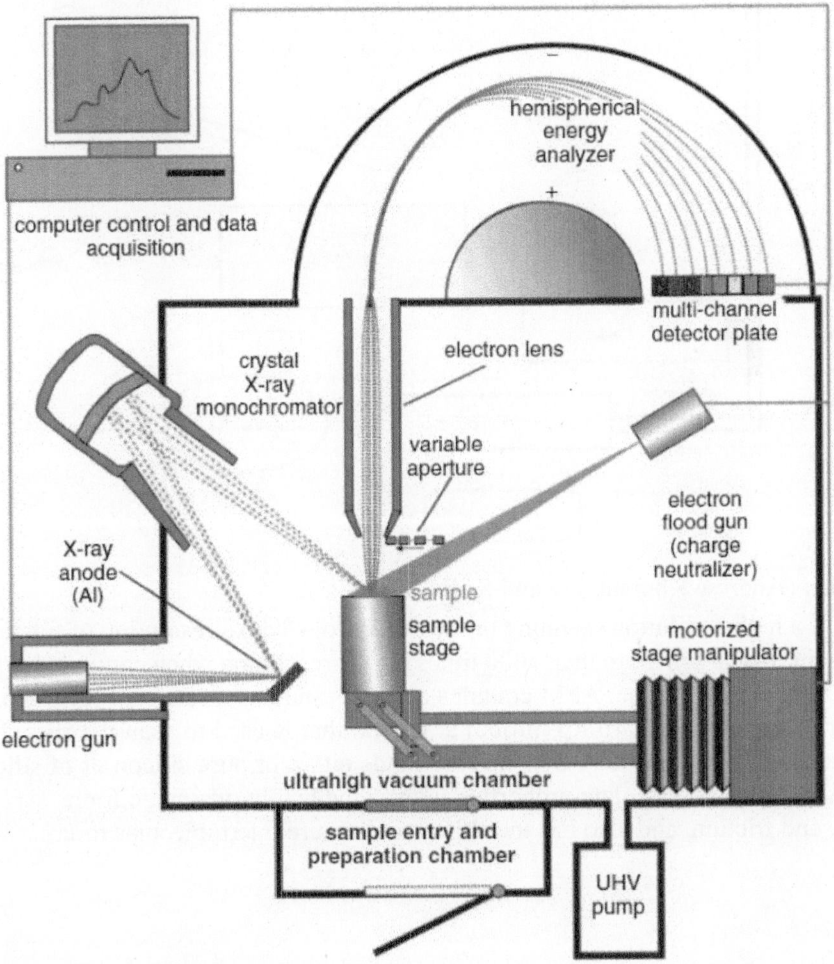

Source:(Fadley, 2010).

It is a technique for analyzing a material's surface chemistry. XPS can measure elemental composition as well as the chemical and electronic state of the atoms within a material.

Atomic force microscope (AFM):

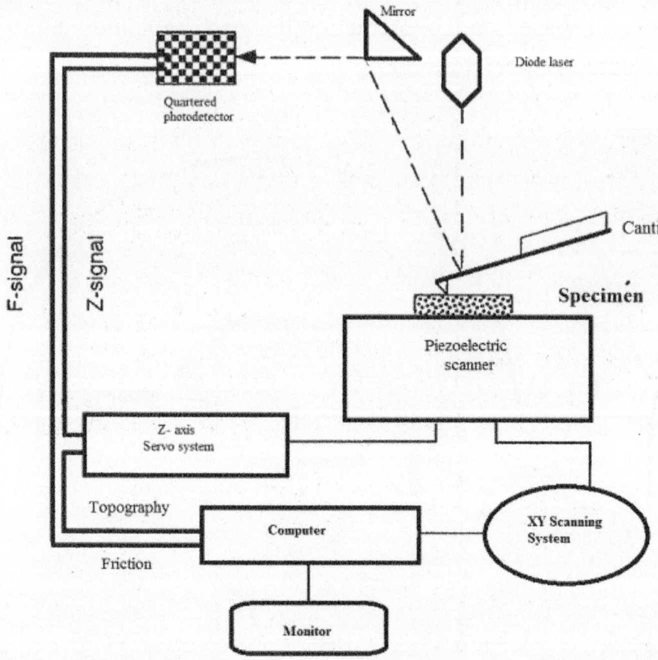

Source: (Andrew Khurshudov and Koji Kato 1995).

AFM is a high-resolution scanning probe microscope with a resolution of fractions of Angstrom, which has more than 1000 times higher resolution when compared to the classical optical microscope. AFM contains a micro- and nanoscale cantilever with a silicon or silicon nitrite sharp tip (probe) at its end that is used to scan the specimen surface at the nanoscale level. AFM tips are made either of pure silicon or of silicon nitride. AFM helps to calculate properties such as the height, magnetic force, surface potential, and friction, and also has the ability to measure intermolecular forces.

5

Abbreviations

S.No.	Acronym and Initialism	Explanation
1.	AAS	Atomic Absorption Spectrometers
2.	ADME	Absorption, Distribution, Metabolism, and Excretion
3.	AES	Atomic Emission Spectrometers
4.	AES	Auger Electron Spectroscopy
5.	AFM	Atomic Force Microscope
6.	AI	Artificial Intelligence
7.	ALD	Atomic Layer Deposition
8.	ALU	Arithmetic Logic Unit
9.	APCVD	Atmospheric Pressure Chemical Vapor Deposition
10.	APD	Avalanche Photo Diode
11.	APT	Atom Probe Tomography
12.	ATP	Adenosine triphosphate
13.	ATR	Attenuated Total Reflectance
14.	AQI	Air Quality Index
15.	BEEM	Ballistic Electron Emission Microscopy
16.	BEES	Ballistic Electron Emission Spectroscopy
17.	BET	Brunauer-Emmett-Teller
18.	Bio-AFM	Biological Atomic Force Microscopes
19.	BPL	Beam Pen Lithography
20.	BTEX	Benzene, Toluene, Ethylbenzene, Xylene
21.	CAU	Carbon Absorption Unit
22.	CDE	Coating and Deposition Equipment
23.	CFC	Chlorofluorocarbons
24.	CFL	Compact Fluorescent Lamp
25.	CFM	Chemical Force Microscopy
26.	CFM	Contact Force Microscopy
27.	CFRP	Carbon Fiber Reinforced Plastics
28.	CHA	Concentric Hemispherical Analyzer

29.	CMA	Cylindrical Mirror Analyzer
30.	CMOS	Complementary metal–oxide–semiconductor
31.	CN	Cellulose Nanomaterial
32.	CNG	Compressed Natural Gas
33.	CNT	Carbon Nanotubes
34.	CPU	Central Processing Unit
35.	CT	Computed Tomography
36.	CVAFS	Cold vapor atomic fluorescence spectroscopy
37.	CVD	Chemical Vapor Deposition
38.	DDT	Dichlorodiphenyltrichloroethane
39.	DDS	Drug Delivery System
40.	DFM	Dynamic Force Microscopy
41.	DFM	Dissipation Force Microscopy
42.	DFS	Dynamic Force Spectroscopy
43.	DLS	Dynamic Light Scattering
44.	DO	Dissolved Oxygen
45.	DPN	Dip-Pen Nanolithography
46.	DRAM	Dynamic Random Access Memory
47.	DSC	Diffraction Scanning Colorimetry
48.	DWNT	Double Walled Nanotubes
49.	EBIC	Electron-Beam-Induced Current
50.	EBSD	Electron Backscattered Diffraction
51.	ECMS	Electrical Conductivity Measurement Systems
52.	EC SPM	Electrochemical Scanning Probe Microscope
53.	EDXRF	Energy-Dispersive X-Ray Fluorescence
54.	EDS	Energy Dispersive X-Ray Spectroscopy
55.	EDX	Energy Dispersive X-ray
56.	EE	Electron Equivalents
57.	EET	Electronic Energy Transfer
58.	EFM	Electrostatic Force Microscopy
59.	EL	Electro Luminescence
60.	ELISA	Enzyme-Linked Immunosorbent Assay
61.	ENP	Engineered Nanoparticles
62.	EPA	Environmental Protection Agency
63.	ERA	Environmental Risk Assessment
64.	FCR	Flow Chemistry Reactors
65.	FCS	Fluorescence Correlation Spectroscopy
66.	FDA	Food and Drug Administration

67.	FEG	Field Emission Gun
68.	FEM	Field Emission Microscope
69.	FESEM	Field Emission Scanning Electron Microscopy
70.	FET	Field-Effect Transistor
71.	FFM	Frictional Force Microscopy
72.	FIB	Focused Ion Beam
73.	FID	Flame Ionization Detection
74.	FIM	Field Ion Microscope
75.	FITC	Fluorescein Isothiocyanate
76.	FM	Fluorescence Microscopes
77.	FMM	Force Modulation Microscopy
78.	FMM	Force Modulation Mode
79.	FPS	Frequency Power Spectrum
80.	FRET	Fluorescence Resonance Energy Transfer
81.	FTIR	Fourier Transform Infrared Spectroscopy
82.	GC	Gas Chromatography
83.	GC-MS	Gas Chromatography-Mass Spectrometry
84.	GFET	Graphene Field Effect Transistors
85.	GHG	Greenhouse Gases
86.	GIS	Geographic Information System
87.	GISAXS	Grazing-incidence small-angle X-ray scattering
88.	GLC	Gas-Liquid Chromatography
89.	GMO	Genetically Modified Organisms
90.	GPC	Gel Permeation Chromatography
91.	GPS	Global Positioning System
92.	GPT	Gas Permeability Tester
93.	GSC	Gas-Solid Chromatography
94.	GTP	Guanosine-5'-triphosphate
95.	HAP	Hazardous Air Pollutants
96.	HARP	High-Area Rapid Printing
97.	HHW	Household Hazardous Waste
98.	HOPG	Highly Oriented Pyrolytic Graphite
99.	HPC	Hybrid Photon-Counting
100.	HPH	High-Pressure Homogenizer
101.	HPLC	High Performance Liquid Chromatography
102.	HTS	High Throughput Screening
103.	IBL	Ion Beam Lithography
104.	ICP-AES	Inductively Coupled Plasma Atomic Emission Spectroscopy

105.	ICP-MS	Inductively Coupled Plasma Mass Spectrometry
106.	IET	Inelastic Electron Tunneling
107.	IFM	Interfacial Force Microscopy
108.	IH	Induction Heating
109.	IPR	Intellectual Property rights
110.	KPM	Kelvin Probe Microscopy
111.	LAS	Laser Absorption Spectroscopy
112.	LASER	Light Amplification of Stimulated Emission Radiation
113.	LB	Langmuir Blodgett
114.	LCD	Liquid Crystal Display
115.	LE	Liposome Extruder
116.	LED	Light Emitting Diode
117.	LETI	Laboratory for electronics and information technology
118.	LFM	Lateral Force Microscopy
119.	LNP	Lipid Nanoparticle
120.	LOC	Lab-On-Chip
121.	LPCVD	Low Pressure Chemical Vapor Deposition
122.	LSI	Large Scale Integration
123.	LTCC	Low Temperature Co-Fired Ceramics Substrates
124.	MALS	Multi Angle Light Scattering
125.	MBE	Molecular Beam Epitaxy
126.	MCT	Micro Combi Tester
127.	ME	Molecular Electronics
128.	MEMS	Micro Electro Mechanical Systems
129.	MFM	Magnetic Force Microscopy
130.	MFM	Manipulation Force Microscopy
131.	MIMS	Molecular Integrated Microsystems
132.	MMRS	Modular Micro Reaction System
133.	MMT	Montmorillonite
134.	MOF	Metal Organic Framework
135.	MOCVD	Metal-Organic Chemical Vapor Deposition
136.	MOSFET	Metal Oxide Semi-conductor Field Effect Transistor
137.	MP-AES	Microwave Plasma Atomic Emission Spectrometry
138.	MPMS	Magnetic Property Measurement System
139.	MRAM	Magnetic Random Access Memory
140.	MRI	Magnetic Resonance Imaging
141.	MVNT	Multinodular and Vacuolating Neuronal Tumor
142.	MWCNT	Multi Walled Carbon Nanotubes
143.	NEMS	Nano Electro Mechanical Systems

144.	NFOM	Near Field Optical Microscopy
145.	NKI	Nanotechnology Knowledge Infrastructure
146.	NL	Nanolithography
147.	nm	nanometer
148.	NMR	Nuclear Magnetic Resonance spectroscopy
149.	NSI	Nanotechnology Signature Initiative
150.	NSE	Nanoscale Science And Engineering
151.	NSOM	Near Field Scanning Optical Microscopy
152.	NST	Nano Scratch Tester
153.	nZVI	Nano zero-valent iron
154.	NTA	Nanoparticle Tracking Analysis
155.	OLED	Organic Light-Emitting Diode
156.	OSHA	Occupational Safety and Health Administration
157.	PAE	Programmable Atomic Equivalents
158.	PALM	Photo Activated Localization Microscopy
159.	PCR	Polymerase Chain Reaction
160.	PCS	Photon Correlation Spectroscopy
161.	PECVD	Plasma Enhanced Chemical Vapor Deposition
162.	PEMFC	Proton-Exchange Membrane Fuel Cells
163.	PMT	Photo Multiplier Tube
164.	PL	Photoluminescence
165.	PPM	Portable Particulate Monitor
166.	PSD	Position Sensitive Detector
167.	PTA	Particle Tracking Analysis
168.	PTRMS	Proton Transfer Reaction Mass Spectrometry
169.	PVD	Physical Vapor Deposition
170.	QCM	Quartz Crystal Microbalance
171.	QFM	Quasi Noncontact Force Microscopy
172.	QPM	Quantum Proximity Microscopy
173.	RC SFM	Resonance Contact Scanning Force Microscope
174.	RFID	Radio frequency identification
175.	RIE	Reactive Ion Etching
176.	SAM	Scanning Acoustic Microscope
177.	SACVD	Sub-Atmospheric Pressure Chemical Vapor Deposition
178.	SAP	Scanning Atom Probe
179.	SAXS	Small Angle X-Ray Scattering
180.	SCM	Scanning Capacitance Microscopy
181.	SEC	Size Exclusion Chromatography
182.	SEM	Scanning Electron Microscopy

183.	SERS	Surface-Enhanced Raman Spectroscopy
184.	SFC	Super Fluid Chromatography
185.	SFE	Supercritical Fluid Extraction
186.	SFM	Scanning Force Microscopy
187.	SFS	Scanning Force Spectroscopy
188.	SHG	Second-Harmonic Generation
189.	SICM	Scanning Ion-Conductance Microscopy
190.	SIFT-MS	Selected Ion Flow Tube-Mass Spectrometry
191.	SIM	Structured Illumination Microscopy
192.	SNOM	Scanning Near Field Optical Microscopy
193.	SOC	Synthetic Organic Chemical
194.	SPE	Scanning Probe Electrochemistry
195.	SPM	Scanning Probe Microscope
196.	SRAM	Static Random-Access Memory
197.	SRS	Stimulated Raman Scattering
198.	STED	Stimulated Emission Depletion Microscopy
199.	STEM	Scanning Transmission Electron Microscopy
200.	STM	Scanning Tunneling Microscope
201.	STORM	Stochastic Optical Reconstruction Microscopy
202.	STS	Scanning Tunneling Spectroscopy
203.	SVET	Scanning Vibrating Electrode Technique
204.	SWNTs	Single Walled Nanotubes
205.	TA	Thermal Analysis
206.	TCD	Thermal Conductivity Detectors
207.	TD	Thermal Desorption
208.	TDS	Total Dissolved Solids
209.	TEM	Tunneling Electron Microscopy
210.	TERS	Tip-Enhanced Raman Spectroscopy
211.	TGA	Thermo Gravimetric Analyzers
212.	TIRF	Total Internal Reflection Fluorescence
213.	TLC	Thin-Layer Chromatography
214.	TRPS	Tunable Resistive Pulse Sensing
215.	TriFC	Trimolecular Fluorescence Complementation
216.	TRI	Toxics Release Inventory
217.	TSS	Total Suspended Solids
218.	UAFM	Ultrasonic Atomic force microscope
219.	UFM	Ultrasonic Force Microscopy
220.	UHV-CVD	Ultra-High Vacuum Chemical Vapor Deposition
221.	UHV	Ultra High Vacuum

222.	UV	Ultraviolet
223.	UV-Vis	Ultraviolet–visible spectroscopy
224.	VPO	Vapor Pressure Osmometry
225.	VOC	Volatile Organic Compound
226.	WAXS	Wide Angle X-ray Scattering
227.	WBC	White Blood Cells
228.	WDS	Wavelength-Dispersive Spectrometry
229.	WDXRF	Wavelength Dispersive XRF Spectrometer
230.	WLA	Weak Link Approach
231.	XAFS	X-ray Absorption Fine Structure
232.	XAS	X-ray Absorption Spectroscopy
233.	XRD	X-Ray Diffractometers
234.	XRF	X-Ray Fluorescence Analyzer
235.	XPS	X-Ray Photoelectron Spectroscopy
236.	XRM	X-Ray Microscopy Systems
237.	ZPA	Zeta Potential Analyzers

6

Student's IQ Corner

1. Giant Magnetoresistor
   (a) Paramagnetic property (b) Ferromagnetic property
   (c) Super paramagnetic property (d) Dimagnetic property
2. A Quantum dot is a __________ particle
   (a) Single dimensional (b) Double dimensional
   (c) Zero dimensional (d) Multi-dimensional
3. Which one is a semiconductor?
   (a) Tube light (b) Iron Oxide
   (c) Diode (d) Carbon Nano tube
4. Uniform deposition of target material over substrate is
   (a) Chemical vapor deposition (b) Attrition
   (c) Sputtering (d) Micro fabrication
5. The wave length of Infra-red radiation is
   (a) 400 – 700 nm (b) 10 nm – 400 nm
   (c) 0.1 cm – 100 cm (d) 780 nm – 1 mm
6. The visible light wave length is
   (a) 200 – 700 nm (b) 100 – 400 nm
   (c) 380 nm – 700 nm (d) 500 nm – 800 nm
7. Which of thr following is an allotrope of carbon?
   (a) Diamond (b) Copper
   (c) Gold (d) Quartz
8. Identify the "correct" statement
   (a) Surfactant act as a catalyst
   (b) Surfactant act as a reductant
   (c) Surfactant act as a oxidizing agent
   (d) Surfactant act are emulsifiers
9. In TEM, electron movement requires
   (a) Inert gas (b) Nobal gas
   (c) Air (d) Vaccum
10. Father of Bio sensor is _______
    (a) Leland Clark (b) Smalley
    (c) Richard Feynman (d) Buckminster

11. The diameter of virus is ________ nm
    (a) 100 (b) 1000
    (c) 10 (d) 10000
12. XRD exploits
    (a) Boltzmann Law (b) Braggs Law
    (c) Newton's Law (d) Stoke's Law
13. Face cream contains
    (a) Titanium Oxide (b) Cerium Oxide
    (c) Tungsten Oxide (d) Iron Oxide
14. Inert Gas condensation Technique
    (a) Mechanical method (b) Physical method
    (c) Chemical method (d) Thermal method
15. Carbon Nanotube was invented by
    (a) Robert curl (b) Richard Smalley
    (c) Sumio lijima (d) Harold Kroto
16. High energy ball milling is _______
    (a) Mechanical (b) Chemical
    (c) Biological (d) Optical
17. Quantum Dot Size ranges from ________
    (a) 2 – 10 nm (b) 50 – 75 nm
    (c) 75 – 100 nm (d) 100 – 1000 nm
18. The size of the DNA is ________
    (a) 50 nm to 60 nm (b) 20 nm to 30 nm
    (c) 2nm to 6 nm (d) 10 nm to 20 nm
19. The meaning of atom by atom manipulation is ________
    (a) Top down approach (b) Nuclear fusion
    (c) Bottom up approach (d) Electron bombardment of samples
20. A measurement of the charge on a particle surface is called ________
    (a) Electro potential (b) Zeta potential
    (c) Kinetic potential (d) Gamma potential
21. Sputtering is a _________ process.
    (a) Physical (b) Chemical
    (c) Biological (d) Optical
22. Strain is the amount of measure of ________
    (a) Formation (b) Plasticity
    (c) Deformation (d) Adhesion
23. Investigation of light matter interaction at the Nano scale
    (a) CNT (b) Nanotubes
    (c) Carbon (d) Nano optics

24. Sol gel synthesis is also refer to _______
    (a) Wet chemical synthesis (b) Physical synthesis
    (c) Green synthesis (d) Thermal evaporation synthesis
25. Name the rule violated during the synthesis of Nano materials ________
    (a) Harrington rule (b) Hall peltch rule
    (c) Hooke's Law (d) Pascal's Law
26. Electro spinning is a unique approach using electrostatic forces to produce ________
    (a) Nanotube (b) Nano fibre
    (c) Quantum dot (d) Carbon tube
27. The size of an atom is _______
    (a) 0.1 nm (b) 0.5 nm
    (c) 10 nm (d) 5 nm
28. Cytoplasmic synthesis of nanoparticles are _________
    (a) Extra cellular (b) Intra cellular
    (c) Self-assembly (d) Both 2 and 3
29. Scanning Electron microscope can be used to study
    (a) Tomography (b) Magnetic properties
    (c) Topography (d) Thermal property
30. Expand MWCNT _______
    (a) Multi walled carbon Nanotubes
    (b) Multi walled carbon Nanotubes
    (c) Mono walled carbon Nanotubes
    (d) Mono walled carbon Nanotubes
31. Which of the following particles is the smallest?
    (a) Proton (b) Electron
    (c) Neutron (d) All of the above
32. Lithography is __________ method of Nano material synthesis.
    (a) Chemical method (b) Biological method
    (c) Physical method (d) Mechanical method
33. Compound microscope was invented by _____
    (a) Eric Drexler (b) Robert Hooke
    (c) Harold croto (d) Richard smalley
34. Dendrimer is a _________
    (a) Organic polymer (b) Organic amorphous
    (c) Organic monomer (d) Inorganic nanoparticle
35. The type of micelle is known as a normal phase micelle _______
    (a) Oil in water micelle (b) Oil in oil micelle
    (c) Water in water micelle (d) Oil in petrol micelle

36. CNT are held together by _____ force forming a rope like structure
    (a) Electric (b) Thermal
    (c) Conductive (d) Van der walls
37. Conductivity of silicon can be charged through the process calles ________
    (a) Diode (b) Dipping
    (c) Anode (d) Doping
38. Molecular beam Epitaxy process is a ________
    (a) Physical method of NPs Synthesis
    (b) Mechanical method of NPs Synthesis
    (c) Chemical method of NPs Synthesis
    (d) Biological method of NPs Synthesis
39. A technique used to prepare thin film is ___________
    (a) Physical method (b) CVD
    (c) Sol-gel (d) Sputtering
40. The carbon nanotubes have been found to have excellent ________ properties.
    (a) Mechanical (b) Magnetic
    (c) Optical (d) None of the above
41. Which one of the following is inert gas?
    (a) Co_2 (b) Ar
    (c) O_2 (d) Co
42. UV Spectroscopy utilizes which of the following electromagnetic source to measure absorbance of the sample
    (a) UV rays (b) X-rays
    (c) Visible light (d) Both 1 and 2
43. Quantum dots are reffered to in nanotechnology as
    (a) Artificial atoms (b) Natural atoms
    (c) Polymers (d) CNT
44. Single tailed lipids in a ___________
    (a) Bilayer (b) Multilayer
    (c) Single (d) Multiple
45. Who published the engines of creation?
    (a) Eric Drexter (b) Richard P. Feynman
    (c) Alexi Ekimovin (d) Norio Taniguchi
46. Which one of the following is a physical process of NPs Synthesis
    (a) Co – Precipitation (b) Micro emulsion
    (c) Sputtering (d) Sol-gel method
47. The scientist invented the electron microscope is
    (a) Mark read (b) Ernst Ruska
    (c) Lijima (d) Robert curl

48. Wave length of the light produced from the quantum depends on the ________ of the quantum particle.
   (a) Shape (b) Size
   (c) Electron arrangements (d) Material
49. The electron beam is scanned or scattered in the following pattern
   (a) Roster Scan (b) Point Scan
   (c) Random Scan (d) Focused Scan
50. Biological Synthesis of nanoparticle is a ________
   (a) Hybrid approach (b) Top down approach
   (c) Plant based approach (d) Bottom up approach
51. The Nanoparticle that extends the seed storability is _______
   (a) ZnO NPs (b) Gold NPs
   (c) Platinum NPs (d) Nickel NPs
52. CNT is stronger than steel by ________
   (a) 10 times (b) 1000 times
   (c) 100 times (d) 10000 times
53. The widely used safety guidelines towards manufacturing and used of nanomaterial were framed by
   (a) OECD (b) FAO
   (c) UNESCO (d) WHO
54. The size of human RBC is ________
   (a) 5 to 100 (b) 100 to 200
   (c) 2000 to 4000 (d) 6000 to 8000
55. Precursors used in sol-gel process are _____
   (a) Surfactants (b) Carbohydrates
   (c) Metals (d) Inorganic salts and metal alkoxide
56. The Quantum dots absorb ________
   (a) UV lights (b) X rays
   (c) Photons (d) Radio waves
57. Which has the largest magnifying capability?
   (a) Diamond (b) Bucky ball
   (c) Graphite (d) Carbon Nanotubes
58. This famous nanomaterial based item is Lycurgus cup coated with ______
   (a) Au (b) Fe
   (c) Ag (d) Ti
59. Example for 1D Nano material is _______
   (a) Nanotube (b) Nano film
   (c) Nano crystal (d) Quantum dots
60. Graphite is the allotrope of ______
   (a) Oxygen (b) Phosphorus
   (c) Selenium (d) Carbon

61. Copper grid sample holder is use ______
    (a) UV-Vis (b) TEM
    (c) FTIR (d) SEM
62. Which is not a chemical method ______
    (a) Sputtering (b) Polyol
    (c) Sol-gel (d) Electro spinning
63. Wind mill blade is coated with ______
    (a) CNTs (b) Silver NPs
    (c) Gold NPs (d) 2N NPs
64. The energy conversion efficiency of Quantum dots based photo voltaic cell is ______
    (a) 80% (b) 65%
    (c) 30% (d) 20%
65. The nano particle used for Gene delivery is ______
    (a) Gold (b) Silver
    (c) Copper (d) Titanium Oxide
66. ______ is the commonly used source of electron in TEM
    (a) Graphene (b) CNT
    (c) Tungsten (d) Aquatrophs
67. Which of the following is enhanced by using Zinc oxide nanoparticle in Agriculture?
    (a) Germination (b) Flowering
    (c) Wilting (d) Senescence
68. According to Moore's Law the amount of space required to install a transistor on a chip shrinks rough by half for ______
    (a) Every 20 months (b) Every 18 months
    (c) Every 14 months (d) Every 36 months
69. Name the magnetic property observed in Nano material ______
    (a) Super para magnetic (b) Para magnetism
    (c) Dia magnetism (d) Ferro magnetism
70. Tendency of solid material to move slowly or deform permanently under the influence of mechanical stress
    (a) Defets (b) Creep
    (c) Plasticity (d) Grain Boundaries
71. The phase 'There's plenty of room at the bottom" was given by ______
    (a) Norio Taniguchi (b) Richard Feynman
    (c) Lijima (d) Moore
72. Electro spinning machine is used to synthesis this Nano material ______
    (a) Nano particle (b) Emulsion
    (c) Nano emulsion (d) Nano fibre

73. Most of the Semiconductor chips and transistors are created with _____
    (a) Carbon nanotubes (b) Diamond
    (c) Silicon (d) Aluminium
74. The ball used in Ball milling is made of _________
    (a) Zirconium (b) Copper
    (c) Gold (d) Silver
75. Sol-gel processing is ___________ method of Nano material Synthesis.
    (a) Mechanical Method (b) Physical Method
    (c) Chemical Method (d) Thermal Method
76. Quantum dots are made from _________
    (a) Zno (b) Cadmium Selenide
    (c) Tio2 (d) Au
77. Carbon atoms make this type of bond with other cation atoms ____
    (a) Covalent (b) Metallic
    (c) Ionic (d) Hydrogen
78. _________ is the element needed for stability of cell membrane
    (a) Nitrogen (b) Manganese
    (c) Calcium (d) Boron
79. The bio-sensor that measures changes in heat is ________
    (a) Piezo electric (b) Optical
    (c) Calorimetric (d) Electrochemical
80. A micelle is an aggregate of surfactant molecule dispersed in a ______
    (a) Liquid Colloid (b) Gaseous phase
    (c) Liquid – Solid phase (d) Solid- Gas phase

ANSWER KEY

1. (c)	2. (c)	3. (c)	4. (a)	5. (d)	6. (c)	7. (c)	8. (d)
9. (d)	10. (a)	11. (a)	12. (b)	13. (a)	14. (b)	15. (c)	16. (a)
17. (a)	18. (c)	19. (c)	20. (b)	21. (a)	22. (c)	23. (d)	24. (a)
25. (b)	26. (b)	27. (a)	28. (b)	29. (c)	30. (b)	31. (b)	32. (c)
33. (b)	34. (a)	35. (a)	36. (d)	37. (d)	38. (a)	39. (d)	40. (b)
41. (b)	42. (a)	43. (a)	44. (c)	45. (a)	46. (c)	47. (b)	48. (b)
49. (a)	50. (d)	51. (a)	52. (c)	53. (a)	54. (d)	55. (d)	56. (c)
57. (d)	58. (a)	59. (a)	60. (d)	61. (b)	62. (a)	63. (a)	64. (b)
65. (a)	66. (c)	67. (a)	68. (b)	69. (a)	70. (b)	71. (b)	72. (d)
73. (c)	74. (a)	75. (c)	76. (b)	77. (a)	78. (c)	79. (c)	80. (a)

Reference

Al-Bukhaiti, W. Q., Noman, A., Qasim, A. S., & Al-Farga, A. (2017). Gas chromatography: Principles, advantages and applications in food analysis. International Journal of Agriculture Innovations and Research, 6(1), 2319-1473.

Czaplicki, S. (2013). Chromatography in bioactivity analysis of compounds. Column chromatography, 99-122.

Fadley, C. S. (2010). X-ray photoelectron spectroscopy: Progress and perspectives. Journal of Electron Spectroscopy and Related Phenomena, 178, 2-32.

Khadka, P., Ro, J., Kim, H., Kim, I., Kim, J. T., Kim, H., & Lee, J. (2014). Pharmaceutical particle technologies: An approach to improve drug solubility, dissolution and bioavailability. Asian journal of pharmaceutical sciences, 9(6), 304-316.

Lakshmanan, A., & Sharmila Rahale, C. (2020). Insights into Fundamentals & Applications of Nanotechnology. MASU, 1-225.

Lakshmanan, A., & Subramanian, K.S. (2020). Nanotechnology in Agriculture, Energy and Environment. Daya Publishing House, 1-701.

Loh, Z. H., Samanta, A. K., & Heng, P. W. S. (2015). Overview of milling techniques for improving the solubility of poorly water-soluble drugs. Asian journal of pharmaceutical sciences, 10(4), 255-274.

Unapumnuk, K., Keener, T. C., Lu, M., & Khang, S. J. (2006). Pyrolysis behavior of tire-derived fuels at different temperatures and heating rates. Journal of the Air & Waste Management Association, 56(5), 618-627.

Zia, K., Siddiqui, T., Ali, S., Farooq, I., Zafar, M. S., & Khurshid, Z. (2019). Nuclear magnetic resonance spectroscopy for medical and dental applications: a comprehensive review. European journal of dentistry, 13(01), 124-128.

https://www.iinano.org/glossary/

https://www.nanotech-now.com/nanotechnology-glossary-A-C.htm

https://trynano.org/education-resources/nanotechnology-glossary/